Engineering Mechanics

Engineering Mechanics

Charlie Fowler

NY RESEARCH
PRESS
New York

Published by NY Research Press
118-35 Queens Blvd., Suite 400,
Forest Hills, NY 11375, USA
www.nyresearchpress.com

Engineering Mechanics
Charlie Fowler

International Standard Book Number: 978-1-63238-796-7 (Hardback)

Cataloging-in-Publication Data

Engineering mechanics / Charlie Fowler.
p. cm.
Includes bibliographical references and index.
ISBN 978-1-63238-796-7
1. Mechanics, Applied. 2. Mechanical engineering. I. Fowler, Charlie.
TA350 .E54 2020
620.1--dc23

Printed in China.

TABLE OF CONTENTS

PREFACE

This book aims to help a broader range of students by exploring a wide variety of significant topics related to this discipline. It will help students in achieving a higher level of understanding of the subject and excel in their respective fields. This book would not have been possible without the unwavered support of my senior professors who took out the time to provide me feedback and help me with the process. I would also like to thank my family for their patience and support.

The scientific branch that deals with the behavior of objects at rest or motion under the influence of forces is referred to as mechanics. Engineering mechanics focuses on the application of the principles of mechanics in order to solve real-time engineering problems. It can be broadly classified into two branches- statics and dynamics. Statics is concerned with the study of objects at rest that may or may not be under the influence of forces. Dynamics is the branch of mechanics which focuses on the study of objects in motion as well as the forces causing such motion. This book discusses the fundamentals as well as modern approaches of engineering mechanics. It strives to provide a fair idea about this discipline and to help develop a better understanding of the latest advances within this field. Those in search of information to further their knowledge will be greatly assisted by this book.

A brief overview of the book contents is provided below:

Chapter – Introduction

Mechanics is the branch of physics that is concerned with the action of forces and energy on macroscopic objects. Engineering mechanics deals with the application of mechanics on the designs and operations of machines. This chapter delves into the subject of engineering mechanics for an in-depth understanding of it.

Chapter – Applied Mechanics

Some of the fundamental principles and concepts that fall under the domain of applied mechanics are vectors, momentum, Newton's laws of motion, friction, law of restitution, etc. This chapter has been carefully written to provide an easy understanding of these principle and concepts of applied mechanics.

Chapter – Statics and Force

Statics comes under the branch of mechanics that deals with the analysis of force and torque of physical systems in static equilibrium with their environment. Force is the resultant action of push or pull on interaction between objects. The topics elaborated in this chapter will provide an easy understanding of the concepts related to statics and force.

Chapter – Dynamics

Dynamics is one of the branches of mechanics that includes the study of forces and their actions on the motion of objects. It includes rotational dynamics, analytical dynamics, rotor dynamics, structural dynamics, vehicle dynamics, etc. All these aspects associated with dynamics have been carefully analyzed in this chapter.

Chapter – Rigid Body Mechanics

A rigid body refers to a solid body in which deformation is zero or negligible. Rigid body mechanics is the study of the movement of interconnected bodies influenced by the external forces. This chapter has been carefully written to provide an extensive understanding of the subject.

Charlie Fowler

Introduction 1

- **Mechanics**
- **Engineering Mechanics**

Mechanics is the branch of physics that is concerned with the action of forces and energy on macroscopic objects. Engineering mechanics deals with the application of mechanics on the designs and operations of machines. This chapter delves into the subject of engineering mechanics for an in-depth understanding of it.

Mechanics

Mechanics is a science concerned with the motion of bodies under the action of forces, including the special case in which a body remains at rest. Of first concern in the problem of motion are the forces that bodies exert on one another. This leads to the study of such topics as gravitation, electricity, and magnetism, according to the nature of the forces involved. Given the forces, one can seek the manner in which bodies move under the action of forces; this is the subject matter of mechanics proper.

Historically, mechanics was among the first of the exact sciences to be developed. Its internal beauty as a mathematical discipline and its early remarkable success in accounting in quantitative detail for the motions of the Moon, the Earth, and other planetary bodies had enormous influence on philosophical thought and provided impetus for the systematic development of science into the 20th century.

Mechanics may be divided into three branches: statics, which deals with forces acting on and in a body at rest; kinematics, which describes the possible motions of a body or system of bodies; and kinetics, which attempts to explain or predict the motion that will occur in a given situation. Alternatively, mechanics may be divided according to the kind of system studied. The simplest mechanical system is the particle, defined as a body so small that its shape and internal structure are of no consequence in the given problem. More complicated is the motion of a system of two or more particles that exert forces on one another and possibly undergo forces exerted by bodies outside of the system.

The principles of mechanics have been applied to three general realms of phenomena. The motions of such celestial bodies as stars, planets, and satellites can be predicted with great accuracy thousands of years before they occur. (The theory of relativity predicts some deviations from the motion according to classical, or Newtonian, mechanics; however, these are so small as to be observable only with very accurate techniques, except in problems involving all or a large portion of the detectable universe.) As the second realm, ordinary objects on Earth down to microscopic size (moving at speeds much lower than that of light) are properly described by classical mechanics without significant corrections. The engineer who designs bridges or aircraft may use the Newtonian laws of classical mechanics with confidence, even though the forces may be very complicated, and the calculations lack the beautiful simplicity of celestial mechanics. The third realm of phenomena comprises the behaviour of matter and electromagnetic radiation on the atomic and subatomic scale. Although there were some limited early successes in describing the behaviour of atoms in terms of classical mechanics, these phenomena are properly treated in quantum mechanics.

Classical mechanics deals with the motion of bodies under the influence of forces or with the equilibrium of bodies when all forces are balanced. The subject may be thought of as the elaboration and application of basic postulates first enunciated by Isaac Newton in his Philosophiae Naturalis Principia Mathematica, commonly known as the Principia. These postulates, called Newton's laws of motion, are set forth below. They may be used to predict with great precision a wide variety of phenomena ranging from the motion of individual particles to the interactions of highly complex systems.

In the framework of modern physics, classical mechanics can be understood to be an approximation arising out of the more profound laws of quantum mechanics and the theory of relativity. However, that view of the subject's place greatly undervalues its importance in forming the context, language, and intuition of modern science and scientists. Our present-day view of the world and man's place in it is firmly rooted in classical mechanics. Moreover, many ideas and results of classical mechanics survive and play an important part in the new physics.

The central concepts in classical mechanics are force, mass, and motion. Neither force nor mass is very clearly defined by Newton, and both have been the subject of much philosophical speculation since Newton. Both of them are best known by their effects. Mass is a measure of the tendency of a body to resist changes in its state of motion. Forces, on the other hand, accelerate bodies, which is to say, they change the state of motion of bodies to which they are applied. The interplay of these effects is the principal theme of classical mechanics.

Although Newton's laws focus attention on force and mass, three other quantities take on special importance because their total amount never changes. These three quantities are energy, (linear) momentum, and angular momentum. Any one of these can be shifted from one body or system of bodies to another. In addition, energy may change form while associated with a single system, appearing as kinetic energy, the energy of motion; potential energy, the energy of position; heat, or internal energy, associated with the random motions of the atoms or molecules composing any real body; or any combination of the three. Nevertheless, the total energy, momentum, and angular momentum in the universe never changes. This fact is expressed in physics by saying that energy, momentum, and angular momentum are conserved. These three conservation laws arise out of Newton's laws, but Newton himself did not express them.

It is a remarkable fact that, although Newton's laws are no longer considered to be fundamental, nor even exactly correct, the three conservation laws derived from Newton's laws—the conservation of energy, momentum, and angular momentum—remain exactly true even in quantum mechanics and relativity. In fact, in modern physics, force is no longer a central concept, and mass is only one of a number of attributes of matter. Energy, momentum, and angular momentum, however, still firmly hold centre stage. The continuing importance of these ideas inherited from classical mechanics may help to explain why this subject retains such great importance in science today.

Units and Dimensions

Quantities have both dimensions, which are an expression of their fundamental nature, and units, which are chosen by convention to express magnitude or size. For example, a series of events have a certain duration in time. Time is the dimension of the duration. The duration might be expressed as 30 minutes or as half an hour. Minutes and hours are among the units in which time may be expressed. One can compare quantities of the same dimensions, even if they are expressed in different units (an hour is longer than a minute). Quantities of different dimensions cannot be compared with one another.

The fundamental dimensions used in mechanics are time, mass, and length. Symbolically, these are written as t, m, and l, respectively. The study of electromagnetism adds an additional fundamental dimension, electric charge, or q. Other quantities have dimensions compounded of these. For example, speed has the dimensions distance divided by time, which can be written as l/t, and volume has the dimensions distance cubed, or l^3. Some quantities, such as temperature, have units but are not compounded of fundamental dimensions.

There are also important dimensionless numbers in nature, such as the number $\pi = 3.14$. Dimensionless numbers may be constructed as ratios of quantities having the same dimension. Thus, the number π is the ratio of the circumference of a circle (a length) to its diameter (another length). Dimensionless numbers have the advantage that they are always the same, regardless of what set of units is being used.

Governments have traditionally been responsible for establishing and enforcing standard units for the sake of orderly commerce, navigation, science, and, of course, taxation. Today all such units are established by international treaty, revised every few years in light of scientific findings. The units used for most scientific measurements are those designated the International System of Units (Système International d'Unités), or SI for short. They are based on the metric system, first adopted officially by France in 1795. Other units, such as those of the British engineering system, are still in use in some places, but these are now defined in terms of the SI units.

The fundamental unit of length is the metre. A metre used to be defined as the distance between two scratch marks on a metal bar kept in Paris, but it is now much more precisely defined as the distance that light travels in a certain time interval (1/299,792,458 of a second). By contrast, in the British system, units of length have a clear human bias: the foot, the inch (the first joint of the thumb), the yard (distance from nose to outstretched fingertip), and the mile (one thousand standard paces of a Roman legion). Each of these is today defined as some fraction or multiple of a metre (one yard is nearly equal to one metre). In the SI or the metric system, lengths are expressed as decimal fractions or multiples of a metre (a millimetre = one-thousandth of a metre; a centimetre = one-hundredth of a metre; a kilometre = one thousand metres).

Times longer than one second are expressed in the units seconds, minutes, hours, days, weeks, and years. Times shorter than one second are expressed as decimal fractions (a millisecond = one-thousandth of a second, a microsecond = one-millionth of a second, and so on). The fundamental unit of time (i.e., the definition of one second) is today based on the intrinsic properties of certain kinds of atoms (an excitation frequency of the isotope cesium-133).

Units of mass are also defined in a way that is technically sound, but in common usage they are the subject of some confusion because they are easily confused with units of weight, which is a different physical quantity. The weight of an object is the consequence of the Earth's gravity operating on its mass. Thus, the mass of a given object is the same everywhere, but its weight varies slightly if it is moved about the surface of the Earth, and it would change a great deal if it were moved to the surface of another planet. Also, weight and mass do not have the same dimensions (weight has the dimensions ml/t^2). The Constitution of the United States, which calls on the government to establish uniform "weights and measures," is oblivious to this distinction, as are merchants the world over, who measure the weight of bread or produce but sell it in units of kilograms, the SI unit of mass. (The kilogram is equal to 1,000 grams; 1 gram is the mass of 1 cubic centimetre of water—under appropriate conditions of temperature and pressure.)

Motion of a Particle in One Dimension

Uniform Motion

According to Newton's first law (also known as the principle of inertia), a body with no net force acting on it will either remain at rest or continue to move with uniform speed in a straight line, according to its initial condition of motion. In fact, in classical Newtonian mechanics, there is no important distinction between rest and uniform motion in a straight line; they may be regarded as the same state of motion seen by different observers, one moving at the same velocity as the particle, the other moving at constant velocity with respect to the particle.

Although the principle of inertia is the starting point and the fundamental assumption of classical mechanics, it is less than intuitively obvious to the untrained eye. In Aristotelian mechanics, and in ordinary experience, objects that are not being pushed tend to come to rest. The law of inertia was deduced by Galileo from his experiments with balls rolling down inclined planes.

For Galileo, the principle of inertia was fundamental to his central scientific task: he had to explain how it is possible that if Earth is really spinning on its axis and orbiting the Sun we do not sense that motion. The principle of inertia helps to provide the answer: Since we are in motion together with Earth, and our natural tendency is to retain that motion, Earth appears to us to be at rest. Thus, the principle of inertia, far from being a statement of the obvious, was once a central issue of scientific contention. By the time Newton had sorted out all the details, it was possible to account accurately for the small deviations from this picture caused by the fact that the motion of Earth's surface is not uniform motion in a straight line (the effects of rotational motion are discussed below). In the Newtonian formulation, the common observation that bodies that are not pushed tend to come to rest is attributed to the fact that they have unbalanced forces acting on them, such as friction and air resistance.

As has already been stated, a body in motion may be said to have momentum equal to the product of its mass and its velocity. It also has a kind of energy that is due entirely to its motion, called kinetic energy. The kinetic energy of a body of mass m in motion with velocity v is given by;

$$K = \frac{1}{2}mv^2$$

Falling Bodies and Uniformly Accelerated Motion

During the 14th century, the French scholar Nicole Oresme studied the mathematical properties of uniformly accelerated motion. He had little interest in whether that kind of motion could be observed in the realm of actual human existence, but he did discover that, if a particle is uniformly accelerated, its speed increases in direct proportion to time, and the distance it traverses is proportional to the square of the time spent accelerating. Two centuries later, Galileo repeated these same mathematical discoveries (perhaps independently) and, just as important, determined that this kind of motion is actually executed by balls rolling down inclined planes. As the incline of the plane increases, the acceleration increases, but the motion continues to be uniformly accelerated. From this observation, Galileo deduced that a body falling freely in the vertical direction would also have uniform acceleration. Even more remarkably, he demonstrated that, in the absence of air resistance, all bodies would fall with the same constant acceleration regardless of their mass. If the constant acceleration of any body dropped near the surface of Earth is expressed as g, the behaviour of a body dropped from rest at height zo and time t = 0 may be summarized by the following equations:

$$z = z_0 - \frac{1}{2}gt^2,$$

$$v = gt,$$

$$a = g$$

where z is the height of the body above the surface, v is its speed, and a is its acceleration. These equations of motion hold true until the body actually strikes the surface. The value of g is approximately 9.8 metres per second squared $\left(\mathrm{m/s^2}\right)$.

A body of mass m at a height z_0 above the surface may be said to possess a kind of energy purely by virtue of its position. This kind of energy (energy of position) is called potential energy. The gravitational potential energy is given by:

$$U = mgz_0.$$

Technically, it is more correct to say that this potential energy is a property of the Earth-body system rather than a property of the body itself, but this pedantic distinction can be ignored.

As the body falls to height z less than z_0, its potential energy U converts to kinetic energy $K = {}^1/_2 mv^2$. Thus, the speed v of the body at any height z is given by solving the equation,

$$\frac{1}{2}mv^2 + mgz = mgz_0.$$

Equation $\frac{1}{2}mv^2 + mgz = mgz_0$ is an expression of the law of conservation of energy. It says that the sum of kinetic energy, ${}^1/_2 mv^2$, and potential energy, mgz, at any point during the fall, is equal to the total initial energy, mgz_o, before the fall began. Exactly the same dependence of speed on height could be deduced from the kinematic equations $z = z_0 - \frac{1}{2}gt^2$, $v = gt$, and $a = g$ above.

$$\frac{1}{2}mv^2 + mgz = mgz_0.$$

In order to reach the initial height z_o, the body had to be given its initial potential energy by some external agency, such as a person lifting it. The process by which a body or a system obtains mechanical energy from outside of itself is called work. The increase of the energy of the body is equal to the work done on it. Work is equal to force times distance.

The force exerted by Earth's gravity on a body of mass m may be deduced from the observation that the body, if released, will fall with acceleration g. Since force is equal to mass times acceleration, the force of gravity is given by F = mg. To lift the body to height z_o, an equal and opposite (i.e., upward) force must be exerted through a distance z_o. Thus, the work done is

$$W = Fz_0 = mgz_0,$$

which is equal to the potential energy that results.

If work is done by applying a force to a body that is not being acted upon by an opposing force, the body is accelerated. In this case, the work endows the body with kinetic energy rather than potential energy. The energy that the body gains is equal to the work done on it in either case. It should be noted that work, potential energy, and kinetic energy, all being aspects of the same quantity, must all have the dimensions ml^2 / t^2.

Simple Harmonic Oscillations

Consider a mass m held in an equilibrium position by springs, as shown in figure. The mass may be perturbed by displacing it to the right or left. If x is the displacement of the mass from equilibrium, the springs exert a force F proportional to x, such that:

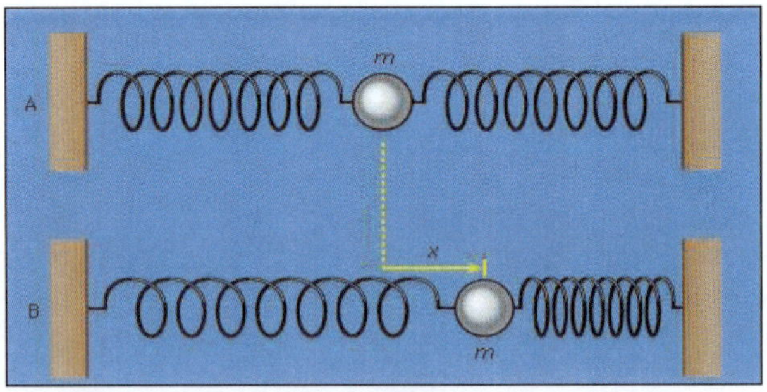

(A) A mass m held in equilibrium by springs. (B) A mass m displaced a distance x.

$$F = -kx,$$

where k is a constant that depends on the stiffness of the springs. Equation $F = -kx$, is called Hooke's law, and the force is called the spring force. If x is positive (displacement to the right), the resulting force is negative (to the left), and vice versa. In other words, the spring force always acts so as to restore mass back toward its equilibrium position. Moreover, the force will produce an acceleration along the x direction given by $a = d^2x/dt^2$. Thus, Newton's second law, F = ma, is applied to this case by substituting $-kx$ for F and d^2x/dt^2 for a, giving $-kx = m(d^2x/dt^2)$. Transposing and dividing by m yields the equation:

$$F = -kx,$$

$$a = \frac{d^2x}{dt^2} = -\frac{k}{m}x.$$

Equation $a = \frac{d^2x}{dt^2} = -\frac{k}{m}x$ gives the derivative—in this case the second derivative—of a quantity x in terms of the quantity itself. Such an equation is called a differential equation, meaning an equation containing derivatives. Much of the ordinary, day-to-day work of theoretical physics consists of solving differential equations. The question is, given equation $a = \frac{d^2x}{dt^2} = -\frac{k}{m}x$, how does x depend on time?

$$a = \frac{d^2x}{dt^2} = -\frac{k}{m}x.$$

The answer is suggested by experience. If the mass is displaced and released, it will oscillate back and forth about its equilibrium position. That is, x should be an oscillating function of t, such as a sine wave or a cosine wave. For example, x might obey a behaviour such as:

$$x = A \cos \omega t.$$

Equation $x = A \cos \omega t$ describes the behaviour sketched graphically in figure. The mass is initially displaced a distance x = A and released at time t = 0. As time goes on, the mass oscillates from A to −A and back to A again in the time it takes ωt to advance by 2π. This time is called T, the period of oscillation, so that ωT = 2π, or T = 2π/ω. The reciprocal of the period, or the frequency f, in oscillations per second, is given by f = 1/T = ω/2π. The quantity ω is called the angular frequency and is expressed in radians per second.

$$x = A \cos \omega t.$$

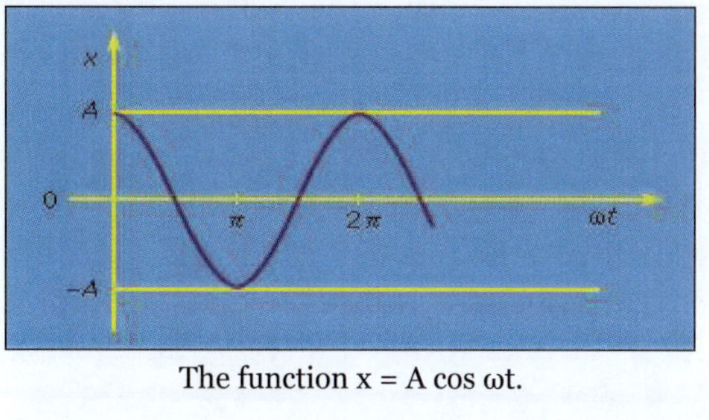

The function x = A cos ωt.

The choice of equation $x = A \cos \omega t$ as a possible kind of behaviour satisfying the differential equation $a = \frac{d^2x}{dt^2} = -\frac{k}{m}x$ can be tested by substituting it into equation $a = \frac{d^2x}{dt^2} = -\frac{k}{m}x$. The first derivative of x with respect to t is;

$$\frac{dx}{dt} = \frac{d}{dt}(A \cos \omega t)$$
$$= -\omega A \sin \omega t.$$

Differentiating a second time gives:

$$\frac{d^2x}{dt^2} = \frac{d}{dt}\left(\frac{dx}{dt}\right)$$
$$= \frac{d}{dt}(-\omega A \sin \omega t)\beta$$
$$= -\omega^2 A \cos \omega t$$
$$= -\omega^2 x.$$

Equation $\frac{d^2x}{dt^2} = \frac{d}{dt}\left(\frac{dx}{dt}\right)$ is the same as equation $a = \frac{d^2x}{dt^2} = -\frac{k}{m}x$ if:

$$\frac{d^2x}{dt^2} = \frac{d}{dt}\left(\frac{dx}{dt}\right)$$
$$= \frac{d}{dt}(-\omega A \sin \omega t)\beta$$
$$= -\omega^2 A \cos \omega t$$
$$= -\omega^2 x.$$

$$\omega^2 = \frac{k}{m}.$$

Thus, subject to this condition, equation $x = A \cos \omega t$ is a correct solution to the differential equation. There are other possible correct guesses (e.g., x = A sin ωt) that differ from this one only in whether the mass is at rest or in motion at the instant t = 0.

The mass, as has been shown, oscillates from A to –A and back again. The speed, given by dx/dt, equation $\frac{dx}{dt} = \frac{d}{dt}(A \cos \omega t) = -\omega A \sin \omega t$, is zero at A and –A, but has its maximum magnitude, equal to ωA, when x is equal to zero. Physically, after the mass is displaced from equilibrium a distance A to the right, the restoring force F pushes the mass back toward its equilibrium position, causing it to accelerate to the left. When it reaches equilibrium, there is no force acting on it at that instant, but it is moving at speed ωA, and its inertia takes it past the equilibrium position. Before it is stopped it reaches position –A, and by this time there is a force acting on it again, pushing it back toward equilibrium.

$$\frac{dx}{dt} = \frac{d}{dt}(A \cos \omega t)$$
$$= -\omega A \sin \omega t.$$

The whole process, known as simple harmonic motion, repeats itself endlessly with a frequency given by equation $\omega^2 = \frac{k}{m}$. Equation $\omega^2 = \frac{k}{m}$ means that the stiffer the springs (i.e., the larger k), the higher the frequency (the faster the oscillations). Making the mass greater has exactly the opposite effect, slowing things down.

$$\omega^2 = \frac{k}{m}.$$

One of the most important features of harmonic motion is the fact that the frequency of the motion, ω (or f), depends only on the mass and the stiffness of the spring. It does not depend on the amplitude A of the motion. If the amplitude is increased, the mass moves faster, but the time required for a complete round trip remains the same. This fact has profound consequences, governing the nature of music and the principle of accurate timekeeping.

The potential energy of a harmonic oscillator, equal to the work an outside agent must do to push the mass from zero to x, is $U = {}^1/_2 kx^2$. Thus, the total initial energy in the situation described above is ${}^1/_2 kA^2$; and since the kinetic energy is always ${}^1/_2 mv^2$, when the mass is at any point x in the oscillation,

$$\frac{1}{2}mv^2 + \frac{1}{2}kx^2 = \frac{1}{2}kA^2.$$

Equation $\frac{1}{2}mv^2 + \frac{1}{2}kx^2 = \frac{1}{2}kA^2$ plays exactly the role for harmonic oscillators that equation $\frac{1}{2}mv^2 + mgz = mgz_0$ does for falling bodies.

$$\frac{1}{2}mv^2 + \frac{1}{2}kx^2 = \frac{1}{2}kA^2.$$

It is quite generally true that harmonic oscillations result from disturbing any body or structure from a state of stable mechanical equilibrium.

Consider a bowl with a marble resting inside, then consider a second, inverted bowl with a marble balanced on top. In both cases, the net force on the marble is zero. The marbles are thus in mechanical equilibrium. However, a small disturbance in the position of the marble balanced on top of the inverted bowl will cause it to roll away and not return. In such a case, the equilibrium is said to be unstable. Conversely, if the marble inside the first bowl is disturbed, gravity acts to push it back toward the bottom of the bowl. The marble inside the bowl (like the mass held by springs in figure) is an example of a body in stable equilibrium. If it is disturbed slightly, it executes harmonic oscillations around the bottom of the bowl rather than rolling away.

This argument may be generalized by a simple mathematical argument. Consider a body or structure in mechanical equilibrium, which, when disturbed by a small amount x, finds a force acting on it that is a function of x, F(x). For small x, such a function may be written generally as a power series in x; i.e.,

$$F(x) = F(0) + ax + bx^2 + ...,$$

where $F(0)$ is the value of $F(x)$ when $x = (0)$, and a and b are constants, independent of x, determined by the nature of the system. The statement that the body is in mechanical equilibrium means that $F(0) = 0$, so that no force is acting on the body when it is undisturbed. Since x is small, x^2 is much smaller; thus, the term bx^2 and all higher powers may be disregarded. This leaves $F(x) = ax$. Now, if a is positive, a disturbance produces a force in the same direction as the disturbance. This was the case when the marble was balanced on top of the inverted bowl. It describes unstable equilibrium. For the system to be stable, a must be negative. Thus, if $a = -k$, where k is some positive constant, equation $F(x) = F(0) + ax + bx^2 + ...,$ becomes $F(x) = -kx$, which is simply Hooke's law, equation $F = -kx$, any system obeying Hooke's law is a harmonic oscillator.

$$F(x) = F(0) + ax + bx^2 + ...,$$

The generality of this argument accounts for the fact that harmonic oscillators are abundantly observed in common experience. For example, any rigid structure will oscillate at many different harmonic frequencies corresponding to different possible distortions of its equilibrium shape. In addition, music may be produced either by disturbing the equilibrium of a stretched wire or fibre (as in the piano and violin), a stretched membrane (e.g., drums), or a rigid bar (the triangle and the xylophone) or by disturbing the density of an enclosed column of air (as in the trumpet and organ). While a fluid such as air is not rigid, its density is an example of a stable system that obeys Hooke's law and may therefore be set into harmonic oscillations.

All music would be quite different from what it is were it not for the general property of harmonic oscillators that the frequency is independent of the amplitude. Thus, instruments yield the same note (frequency) regardless of how loudly they are played (amplitude), and, equally important, the same note persists as the vibrations die away. This same property of harmonic oscillators is the underlying principle of all accurate timekeeping.

The first precise timekeeping mechanism, whose principles of motion were discovered by Galileo, was the simple pendulum. The accuracy of modern timekeeping has been improved dramatically by the introduction of tiny quartz crystals, whose harmonic oscillations generate electrical signals that may be incorporated into miniaturized circuits in clocks and wristwatches. All harmonic oscillators are natural timekeeping devices because they oscillate at intrinsic natural frequencies independent of amplitude. A given number of complete cycles always corresponds to the same elapsed time. Quartz crystal oscillators make more accurate clocks than pendulums do principally because they oscillate many more times per second.

Damped and Forced Oscillations

The simple harmonic oscillations discussed above continue forever, at constant amplitude, oscillating as shown in figure between A and –A. Common experience indicates that real oscillators behave somewhat differently, however. Harmonic oscillations tend to die away as time goes on. This behaviour, called damping of the oscillations, is produced by forces such as friction and viscosity. These forces are known collectively as dissipative forces because they tend to dissipate the potential and kinetic energies of macroscopic bodies into the energy of the chaotic motion of atoms and molecules known as heat.

Friction and viscosity are complicated phenomena whose effects cannot be represented accurately by a general equation. However, for slowly moving bodies, the dissipative forces may be represented by:

$$F_d = -\gamma v,$$

Where v is the speed of the body and γ is a constant coefficient, independent of dynamic quantities such as speed or displacement. Equation $F_d = -\gamma v$, is most easily understood by an argument analogous to that applied to equation $F(x) = F(0) + ax + bx^2 + ...$, above. F_d is written as a sum of powers of v, or $F_d(v) = F_d(0) + av + bv^2 + \cdots$. When the body is at rest (v = 0), no dissipative force is expected because, if there were one, it might set the body into motion. Thus, $F_d(0) = 0$. The next term must be negative since dissipative forces always resist the motion. Thus, $a = -\gamma$ where γ is positive. Since v^2 has the same sign regardless of the direction of the motion, b must equal 0 lest it sometimes contribute a dissipative force in the same direction as the motion. The next term is proportional to v^3, and it and all subsequent terms may be neglected if v is sufficiently small. So, as in equation $F(x) = F(0) + ax + bx^2 + ...$, the power series is reduced to a single term, in this case $F_d = -\gamma v$.

$$F_d = -\gamma v,$$

To find the effect of a dissipative force on a harmonic oscillator, a new differential equation must be solved. The net force, or mass times acceleration, written as md^2x/dt^2, is set equal to the sum of the Hooke's law force, $-kx$, and the dissipative force, $-\gamma v = -\gamma dx / dt$ Dividing by m yields:

$$\frac{d^2x}{dt^2} = -\frac{k}{m}x - \frac{\gamma}{m}\frac{dx}{dt}.$$

The general solution to equation $\frac{d^2x}{dt^2} = -\frac{k}{m}x - \frac{\gamma}{m}\frac{dx}{dt}$ is given in the form $x = Ce^{-\gamma}t^{/2}m\cos(\omega t + \theta_0)$, where C and θ_0 are arbitrary constants determined by the initial conditions. This motion, for the case in which $\theta_0 = 0$, is illustrated in figure. As expected, the harmonic oscillations die out with time. The amplitude of the oscillations is bounded by an exponentially decreasing function of time (the dashed curves). The characteristic decay time (after which the oscillations are smaller by $1/e$, where e is the base of the natural logarithms e = 2.718...) is equal to $2m / \gamma$ The frequency of the oscillations is given by:

$$\frac{d^2x}{dt^2} = -\frac{k}{m}x - \frac{\gamma}{m}\frac{dx}{dt}.$$

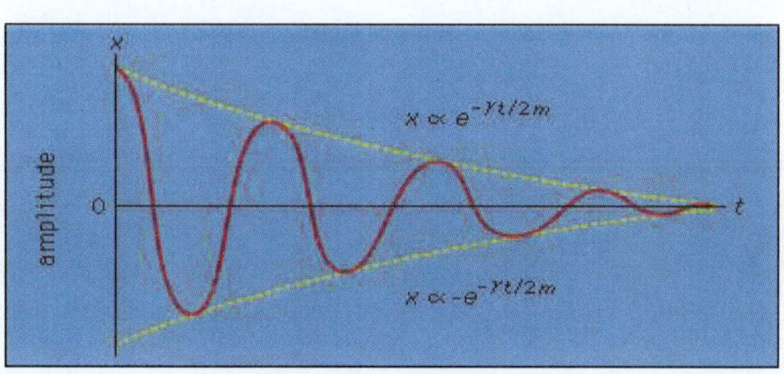

Damped oscillations.

$$\omega^2 = \frac{k}{m} - \frac{\gamma^2}{4m^2}.$$

Importantly, this frequency does not change as the oscillations decay.

Equation $\omega^2 = \frac{k}{m} - \frac{\gamma^2}{4m^2}$ shows that it is possible, by proper choice of γ, to turn a harmonic oscillator into a system that does not oscillate at all—that is, a system whose natural frequency is $\omega = 0$. Such a system is said to be critically damped. For example, the springs that suspend the body of an automobile cause it to be a natural harmonic oscillator. The shock absorbers of the auto are devices that seek to add just enough dissipative force to make the assembly critically damped. In this way, the passengers need not go through numerous oscillations after each bump in the road.

$$\omega^2 = \frac{k}{m} - \frac{\gamma^2}{4m^2}.$$

A simple disturbance can set a harmonic oscillator into motion. Repeated disturbances can increase the amplitude of the oscillations if they are applied in synchrony with the natural frequency. Even a very small disturbance, repeated periodically at just the right frequency, can cause a very large amplitude motion to build up. This phenomenon is known as resonance.

Periodically forced oscillations may be represented mathematically by adding a term of the form $a_0 \sin \omega t$ to the right-hand side of equation $\frac{d^2x}{dt^2} = -\frac{k}{m}x - \frac{\gamma}{m}\frac{dx}{dt}$. This term describes a force applied at frequency ω, with amplitude ma_0. The result of applying such a force is to create a kind of motion that does not need to decay with time, since the energy lost to dissipative processes is replaced, over the course of each cycle, by the driving force. The amplitude of the motion depends on how close the driving frequency ω is to the natural frequency ω_0 of the oscillator. Interestingly, even though dissipation is present, ω_0 is not given by equation $\omega^2 = \frac{k}{m} - \frac{\gamma^2}{4m^2}$ but rather by equation $\omega^2 = \frac{k}{m}$: $\omega^2_0 = k/m$. In a graph of the amplitude of the steady state motion (i.e., long after the driving force has begun to be applied), the maximum amplitude occurs as expected at $\omega = \omega_0$. The height and width of the resonance curve are governed by the damping coefficient γ. If there were no damping, the maximum amplitude would be infinite. Because small disturbances at every possible frequency are always present in the natural world, every rigid structure would shake itself to pieces if not for the presence of internal damping.

Resonances are not uncommon in the world of familiar experience. For example, cars often rattle at certain engine speeds, and windows sometimes rattle when an airplane flies by. Resonance is particularly important in music. For example, the sound box of a violin does its job well if it has a natural frequency of oscillation that responds resonantly to each musical note. Very strong resonances to certain notes—called "wolf notes" by musicians—occur in cheap violins and are much to be avoided. Sometimes, a glass may be broken by a singer as a result of its resonant response to a particular musical note.

Motion of a Particle in Two or More Dimensions

Projectile Motion

Galileo was quoted above pointing out with some detectable pride that none before him had realized that the curved path followed by a missile or projectile is a parabola. He had arrived at his conclusion by realizing that a body undergoing ballistic motion executes, quite independently, the motion of a freely falling body in the vertical direction and inertial motion in the horizontal direction. These considerations, and terms such as ballistic and projectile, apply to a body that, once launched, is acted upon by no force other than Earth's gravity.

Projectile motion may be thought of as an example of motion in space—that is to say, of three-dimensional motion rather than motion along a line, or one-dimensional motion. In a suitably defined system of Cartesian coordinates, the position of the projectile at any instant may be specified by giving the values of its three coordinates, $x(t)$, $y(t)$, and $z(t)$. By generally accepted convention, $z(t)$ is used to describe the vertical direction. To a very good approximation, the motion is confined to a single vertical plane, so that for any single projectile it is possible to choose a coordinate system such that the motion is two-dimensional [say, $x(t)$ and $z(t)$] rather than three-dimensional [$x(t)$, $y(t)$, and $z(t)$].

Consider a body whose vertical motion obeys equation $z = z_0 - \frac{1}{2}gt^2$, Galileo's law of falling bodies, which states $z = z_0 - {}^1/_2 gt^2$ while, at the same time, moving horizontally at a constant speed vx in accordance with Galileo's law of inertia. The body's horizontal motion is thus described by $x(t) = vxt$, which may be written in the form $t = x/vx$. Using this result to eliminate t from equation $z = z_0 - \frac{1}{2}gt^2$ gives $z = z_0 - {}^1/_2 g(1/vx)^2 x^2$. This latter is the equation of the trajectory of a projectile in the z–x plane, fired horizontally from an initial height z_0. It has the general form:

$$z = a + bx^2,$$

where a and b are constants. Equation $z = a + bx^2$, may be recognized to describe a parabola, just as Galileo claimed. The parabolic shape of the trajectory is preserved even if the motion has an initial component of velocity in the vertical direction.

$$z = a + bx^2,$$

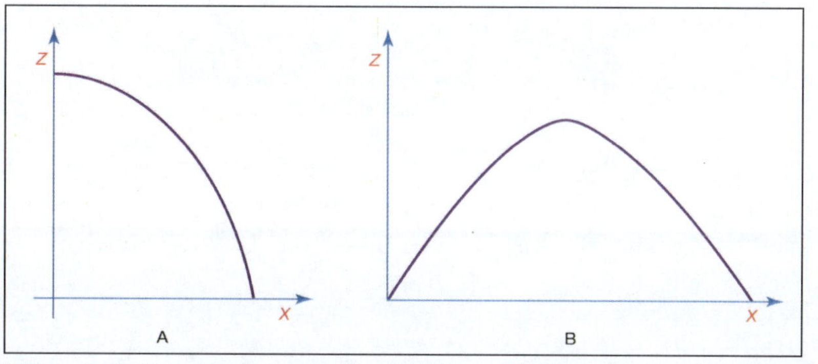

(A) The parabolic path of a projectile. (B) The parabolic path of a projectile with an initial upward component of velocity.

Energy is conserved in projectile motion. The potential energy U(z) of the projectile is given by U(z) = mgz. The kinetic energy K is given by $K = {}^1/_2 mv^2$, where v2 is equal to the sum of the squares of the vertical and horizontal components of velocity, or $v^2 = v^2x + v^2z$.

The effects of air resistance (to say nothing of wind and other more complicated phenomena) have been neglected. These effects are seldom actually negligible. They are most nearly so for bodies that are heavy and slow-moving.

Motion of a Pendulum

According to legend, Galileo discovered the principle of the pendulum while attending mass at the Duomo (cathedral) located in the Piazza del Duomo of Pisa, Italy. A lamp hung from the ceiling by a cable and, having just been lit, was swaying back and forth. Galileo realized that each complete cycle of the lamp took the same amount of time, compared to his own pulse, even though the amplitude of each swing was smaller than the last. As has already been shown, this property is common to all harmonic oscillators, and, indeed, Galileo's discovery led directly to the invention of the first accurate mechanical clocks. Galileo was also able to show that the period of oscillation of a simple pendulum is proportional to the square root of its length and does not depend on its mass.

A simple pendulum is sketched in figure. A bob of mass M is suspended by a massless cable or bar of length L from a point about which it pivots freely. The angle between the cable and the vertical is called θ. The force of gravity acting on the mass M, always equal to −Mg in the vertical direction, is a vector that may be resolved into two components, one that acts ineffectually along the cable and another, perpendicular to the cable, that tends to restore the bob to its equilibrium position directly below the point of suspension. This latter component is given by:

$$F = -Mg \sin\theta.$$

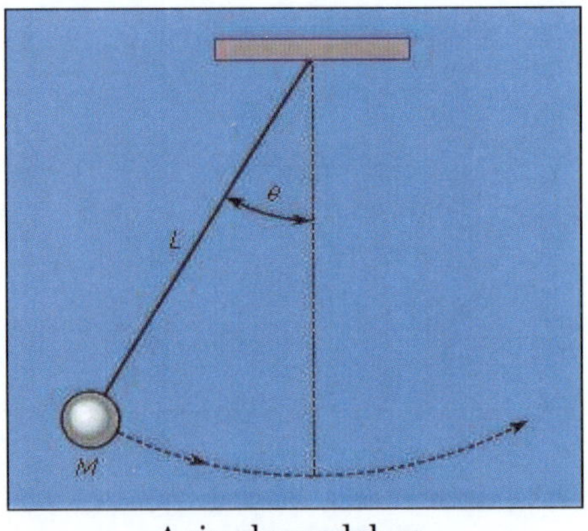

A simple pendulum.

The bob is constrained by the cable to swing through an arc that is actually a segment of a circle of radius L. If the cable is displaced through an angle θ, the bob moves a distance Lθ along its arc (θ must be expressed in radians for this form to be correct). Thus, Newton's second law may be written:

$$F = Ma = M\frac{d^2(L\theta)}{dt^2}.$$

Equating equation $F = -Mg\sin\theta$ to equation $F = Ma = M\frac{d^2(L\theta)}{dt^2}$, one sees immediately that the mass M will drop out of the resulting equation. The simple pendulum is an example of a falling body, and its dynamics do not depend on its mass for exactly the same reason that the acceleration of a falling body does not depend on its mass: both the force of gravity and the inertia of the body are proportional to the same mass, and the effects cancel one another. The equation that results (after extracting the constant L from the derivative and dividing both sides by L) is:

$$F = -Mg\sin\theta.$$

$$\frac{d^2\theta}{dt^2} = -\frac{g}{L}\sin\theta.$$

If the angle θ is sufficiently small, equation $\frac{d^2\theta}{dt^2} = -\frac{g}{L}\sin\theta$ may be rewritten in a form that is both more familiar and more amenable to solution. Figure shows a segment of a circle of radius L. A radius vector at angle θ, as shown, locates a point on the circle displaced a distance Lθ along the arc. It is clear from the geometry that L sin θ and Lθ are very nearly equal for small θ. It follows then that sin θ and θ are also very nearly equal for small θ. Thus, if the analysis is restricted to small angles, then sin θ may be replaced by θ in equation $\frac{d^2\theta}{dt^2} = -\frac{g}{L}\sin\theta$ to obtain:

$$\frac{d^2\theta}{dt^2} = -\frac{g}{L}\sin\theta.$$

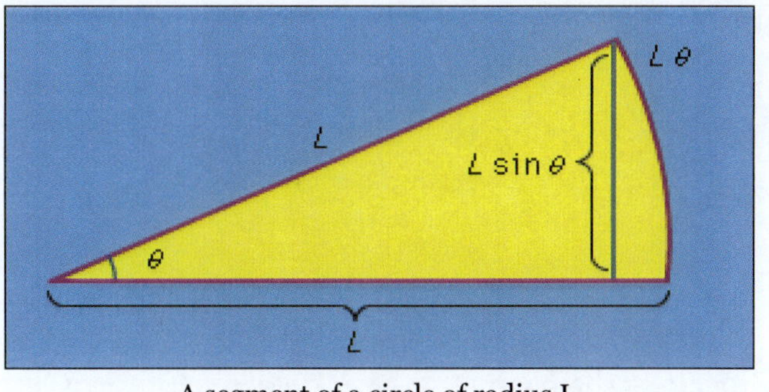

A segment of a circle of radius L.

$$\frac{d^2\theta}{dt^2} = -\frac{g}{L}\theta.$$

Equation $\frac{d^2\theta}{dt^2} = -\frac{g}{L}\theta$ should be compared with equation $a = \frac{d^2x}{dt^2} = -\frac{k}{m}x$: $d^2x/dt^2 = -(k/m)x$. In the first case, the dynamic variable (meaning the quantity that changes with time) is θ, in the second case it is *x*. In both cases, the second derivative of the dynamic variable with respect to time is equal to the variable itself multiplied by a negative constant. The equations are therefore mathematically identical and have the same solution—i.e., equation $x = A\cos\omega t$, or $\theta = A\cos\omega t$. In the case of the pendulum, the frequency of the oscillations is given by the constant in equation $\frac{d^2\theta}{dt^2} = -\frac{g}{L}\theta$, or $\omega^2 = g/L$. The period of oscillation, $T = 2\pi/\omega$, is therefore,

$$\frac{d^2\theta}{dt^2} = -\frac{g}{L}\theta.$$

$$T = 2\pi\sqrt{\frac{L}{g}}.$$

Just as Galileo concluded, the period is independent of the mass and proportional to the square root of the length.

As with most problems in physics, this discussion of the pendulum has involved a number of simplifications and approximations. Most obviously, sin θ was replaced by θ to obtain equation $\frac{d^2\theta}{dt^2} = -\frac{g}{L}\theta$. This approximation is surprisingly accurate. For example, at a not-very-small angle of 17.2°, corresponding to 0.300 radian, sin θ is equal to 0.296, an error of less than 2 percent. For smaller angles, of course, the error is appreciably smaller.

The problem was also treated as if all the mass of the pendulum were concentrated at a point at the end of the cable. This approximation assumes that the mass of the bob at the end of the cable is much larger than that of the cable and that the physical size of the bob is small compared with the length of the cable. When these approximations are not sufficient, one must take into account the way in which mass is distributed in the cable and bob. This is called the physical pendulum, as opposed to the idealized model of the simple pendulum. Significantly, the period of a physical pendulum does not depend on its total mass either.

The effects of friction, air resistance, and the like have also been ignored. These dissipative forces have the same effects on the pendulum as they do on any other kind of harmonic oscillator, They cause the amplitude of a freely swinging pendulum to grow smaller on successive swings. Conversely, in order to keep a pendulum clock going, a mechanism is needed to restore the energy lost to dissipative forces.

Circular Motion

Consider a particle moving along the perimeter of a circle at a uniform rate, such that it makes one complete revolution every hour. To describe the motion mathematically, a vector is constructed from the centre of the circle to the particle. The vector then makes one complete revolution every hour. In other words, the vector behaves exactly like the large hand on a wristwatch, an arrow of fixed length that makes one complete revolution every hour. The motion of the point of the vector is an example of uniform circular motion, and the period T of the motion is equal to one hour (T = 1 h). The arrow sweeps out an angle of 2π radians (one complete circle) per hour. This rate is called the angular frequency and is written $\omega = 2\pi\ h^{-1}$. Quite generally, for uniform circular motion at any rate,

$$T = \frac{2\pi}{\omega}.$$

These definitions and relations are the same as they are for harmonic motion.

Consider a coordinate system, as shown in figure, with the circle centred at the origin. At any instant of time, the position of the particle may be specified by giving the radius r of the circle and the angle θ between the position vector and the x-axis. Although r is constant, θ increases uniformly with time t, such that $\theta = \omega t$, or $d\theta / dt = \omega$, where ω is the angular frequency in equation $T = \frac{2\pi}{\omega}$. Contrary to the case of the wristwatch, however, ω is positive by convention when the rotation is in the counter clockwise sense. The vector r has x and y components given by:

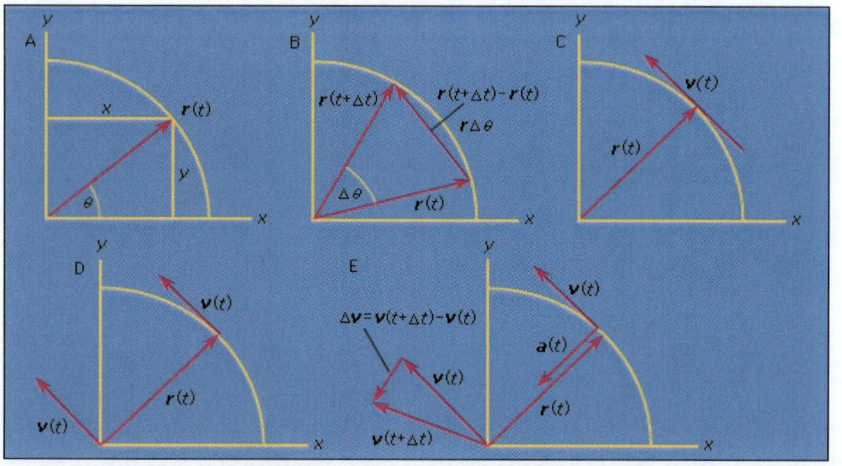

(A) A coordinate system to describe uniform circular motion. (B) The distance traveled in time Δt by a particle undergoing uniform circular motion. (C) The instantaneous velocity of the particle. (D) The velocity vector v undergoes uniform circular motion at the same angular frequency as the particle. (E) The acceleration vector of the particle.

$$x = r\cos\theta = r\cos\omega t,$$

$$y = r\sin\theta = r\sin\omega t.$$

One meaning of equations $x = r\cos\theta = r\cos\omega t$, and $y = r\sin\theta = r\sin\omega t$ is that, when a particle undergoes uniform circular motion, its x and y components each undergo simple harmonic motion. They are, however, not in phase with one another: at the instant when x has its maximum amplitude (say, at θ = 0), y has zero amplitude, and vice versa.

$$x = r\cos\theta = r\cos\omega t,$$

$$y = r\sin\theta = r\sin\omega t.$$

In a short time, Δt, the particle moves rΔθ along the circumference of the circle, as shown in figure. The average speed of the particle is thus given by:

$$\vec{v} = r\frac{\Delta\theta}{\Delta t}.$$

The average velocity of the particle is a vector given by:

$$\vec{v} = \frac{r(t+\Delta t) - r(t)}{\Delta t}.$$

This operation of vector subtraction is indicated in figure. It yields a vector that is nearly perpendicular to $r(t)$ and $r(t + \Delta t)$. Indeed, the instantaneous velocity, found by allowing Δt to shrink to zero, is a vector v that is perpendicular to r at every instant and whose magnitude is:

$$|v| = r\frac{d\theta}{dt} = r\omega.$$

The relationship between r and v is shown in figure. It means that the particle's instantaneous velocity is always tangent to the circle.

Notice that, just as the position vector r may be described in terms of the components x and y given by equations $y = r\ \sin\theta = r\sin\omega t$ and $y = r\ \sin\theta = r\sin\omega t$, the velocity vector v may be described in terms of its projections on the x and y axes, given by:

$$v_x = \frac{dx}{dt} = -r\omega\ \sin\ \omega t,$$

$$v_y = \frac{dy}{dt} = -r\omega\ \cos\ \omega t,$$

Imagine a new coordinate system, in which a vector of length ωr extends from the origin and points at all times in the same direction as v. This construction is shown in figure. Each time the particle sweeps out a complete circle, this vector also sweeps out a complete circle. In fact, its point is executing uniform circular motion at the same angular frequency as the particle itself. Because vectors have magnitude and direction, but not position in space, the vector that has been constructed is the velocity v. The velocity of the particle is itself undergoing uniform circular motion at angular frequency ω.

Although the speed of the particle is constant, the particle is nevertheless accelerated, because its velocity is constantly changing direction. The acceleration a is given by:

$$a = \frac{dv}{dt}.$$

Since v is a vector of length $r\omega$ undergoing uniform circular motion, equations $\vec{v} = r\frac{\Delta\theta}{\Delta t}$. and $\vec{v} = \frac{r(t+\Delta t) - r(t)}{\Delta t}$. may be repeated, as illustrated in figure, giving:

$$\vec{v} = r\frac{\Delta\theta}{\Delta t}.$$

$$\vec{v} = \frac{r(t+\Delta t) - r(t)}{\Delta t}.$$

$$\vec{a} = r\omega\frac{\Delta\theta}{\Delta t}$$

$$\vec{a} = \frac{v(t+\Delta t) - v(t)}{\Delta t}.$$

Thus, one may conclude that the instantaneous acceleration is always perpendicular to v and its magnitude is:

$$|a| = r\omega \frac{d\theta}{dt} = r\omega^2.$$

Since v is perpendicular to r, and a is perpendicular to v, the vector a is rotated 180° with respect to r. In other words, the acceleration is parallel to r but in the opposite direction. The same conclusion may be reached by realizing that a has x and y components given by,

$$a_x = \frac{dv_x}{dt} = -r\omega^2 \cos\omega t,$$

$$a_y = \frac{dv_y}{dt} = -r\omega^2 \sin\omega t,$$

similar to equations $v_x = \frac{dx}{dt} = -r\omega \sin \omega t$, and $v_y = \frac{dy}{dt} = -r\omega \cos \omega t$. When equations $a_x = \frac{dv_x}{dt}$, $= -r\omega^2 \cos\omega t$, and $a_y = \frac{dv_y}{dt} = -r\omega^2 \sin\omega t$, are compared with equations $x = r \cos\theta = r \cos \omega t$, and $y = r \sin\theta = r\sin\omega t$ for x and y, it is clear that the components of a are just those of r multiplied by $-\omega^2$, so that $a = -\omega^2 r$. This acceleration is called the centripetal acceleration, meaning that it is inward, pointing along the radius vector toward the centre of the circle. It is sometimes useful to express the centripetal acceleration in terms of the speed v. Using $v = \omega r$, one can write:

$$v_x = \frac{dx}{dt} = -r\omega \sin \omega t,$$

$$v_y = \frac{dy}{dt} = -r\omega \cos \omega t,$$

$$a_x = \frac{dv_x}{dt} = -r\omega^2 \cos\omega t,$$

$$a_y = \frac{dv_y}{dt} = -r\omega^2 \sin\omega t,$$

$$a = -\frac{v^2}{r}.$$

Circular Orbits

A body in uniform circular motion undergoes at all times a centripetal acceleration given by equation. According to Newton's second law, a force is required to produce this acceleration. In the case of an orbiting planet, the force is gravity. The situation is illustrated in figure. The gravitational attraction of the Sun is an inward (centripetal) force acting on Earth. This force produces the centripetal acceleration of the orbital motion.

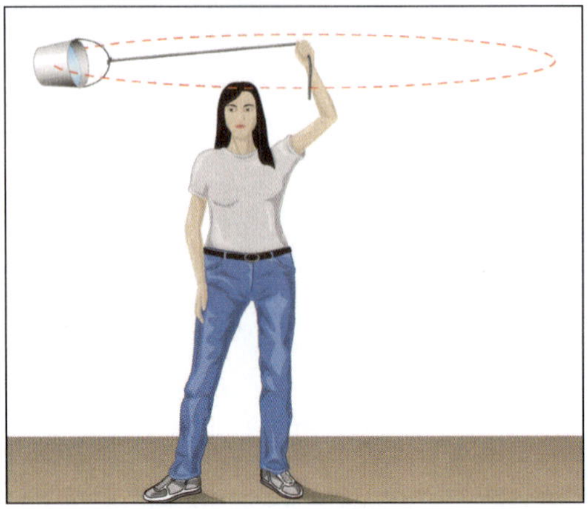

Centripetal force The bucket experiences a centripetal force directed along the string toward the centre of the circle.

Before these ideas are expressed quantitatively, an understanding of why a force is needed to maintain a body in an orbit of constant speed is useful. The reason is that, at each instant, the velocity of the planet is tangent to the orbit. In the absence of gravity, the planet would obey the law of inertia (Newton's first law) and fly off in a straight line in the direction of the velocity at constant speed. The force of gravity serves to overcome the inertial tendency of the planet, thereby keeping it in orbit.

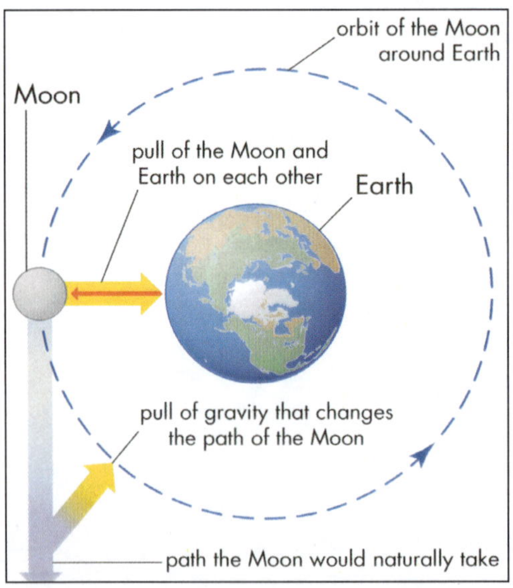

Effects of gravity on Earth and the Moon.

The gravitational force between two bodies such as the Sun and Earth is given by,

$$F = -G\frac{M_S M_E}{r^2}.$$

where M_S and M_E are the masses of the Sun and Earth, respectively, r is the distance between their centres, and G is a universal constant equal to 6.674×10^{-11} Nm^2/kg^2 (Newton metres squared per kilogram squared). The force acts along the direction connecting the two bodies (i.e., along the radius vector of the uniform circular motion), and the minus sign signifies that the force is attractive, acting to pull Earth toward the Sun.

To an observer on the surface of Earth, the planet appears to be at rest at (approximately) a constant distance from the Sun. It would appear to the observer, therefore, that any force (such as the Sun's gravity) acting on Earth must be balanced by an equal and opposite force that keeps Earth in equilibrium. In other words, if gravity is trying to pull Earth into the Sun, some opposing force must be present to prevent that from happening. In reality, no such force exists. Earth is in freely accelerated motion caused by an unbalanced force. The apparent force, known in mechanics as a pseudoforce, is due to the fact that the observer is actually in accelerated motion. In the case of orbital motion, the outward pseudoforce that balances gravity is called the centrifugal force.

For a uniform circular orbit, gravity produces an inward acceleration given by equation $a = -\frac{v^2}{r}$, $a = -v^2 / r$. The pseudoforce f needed to balance this acceleration is just equal to the mass of Earth times an equal and opposite acceleration, or $f = M_E v^2 / r$. The earthbound observer then believes that there is no net force acting on the planet—i.e., that $F + f = 0$, where F is the force of gravity given by equation $F = -G\frac{M_S M_E}{r^2}$. Combining these equations yields a relation between the speed v of a planet and its distance r from the Sun:

$$F = -G\frac{M_S M_E}{r^2}.$$

$$v^2 = G\frac{M_S}{r}.$$

It should be noted that the speed does not depend on the mass of the planet. This occurs for exactly the same reason that all bodies fall toward Earth with the same acceleration and that the period of a pendulum is independent of its mass. An orbiting planet is in fact a freely falling body.

Equation $v^2 = G\frac{M_S}{r}$ is a special case (for circular orbits) of Kepler's third law, Using the fact that $v = 2\pi r / T$, where $2\pi r$ is the circumference of the orbit and T is the time to make a complete orbit (i.e., T is one year in the life of the planet), it is easy to show that $T^2 = \left(4\pi^2 / GM_S\right) r^3$. This relation also may be applied to satellites in circular orbit around Earth (in which case, *ME* must be substituted for M_S) or in orbit around any other central body.

$$v^2 = G\frac{M_S}{r}.$$

Angular Momentum and Torque

A particle of mass m and velocity v has linear momentum $p = mv$. The particle may also have angular momentum L with respect to a given point in space. If r is the vector from the point to the particle, then:

$$L = r \times p.$$

Notice that angular momentum is always a vector perpendicular to the plane defined by the vectors r and p (or v). For example, if the particle (or a planet) is in a circular orbit, its angular momentum with respect to the centre of the circle is perpendicular to the plane of the orbit and in the direction given by the vector cross product right-hand rule, as shown in figure. Moreover, since in the case of a circular orbit, r is perpendicular to p (or v), the magnitude of L is simply:

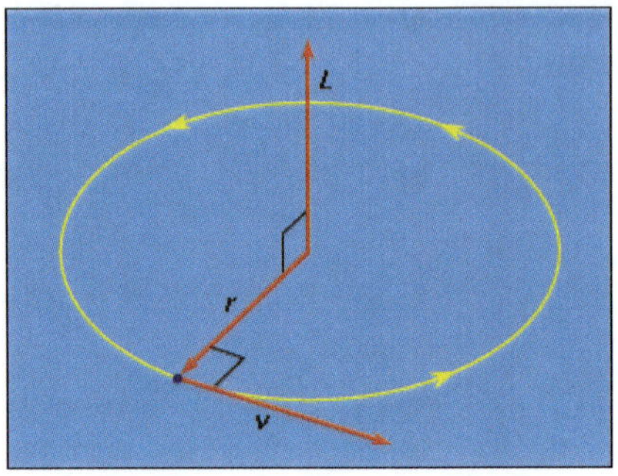

The angular momentum L of a particle traveling in a circular orbit.

$$L = rp = mvr.$$

The significance of angular momentum arises from its derivative with respect to time,

$$\frac{dL}{dt} = \frac{d}{dt}(r \times p) = m\frac{d}{dt}(r \times v),$$

where p has been replaced by mv and the constant m has been factored out. Using the product rule of differential calculus,

$$\frac{d}{dt}(r \times v) = \frac{dr}{dt} \times v + r \times \frac{dv}{dt}.$$

In the first term on the right-hand side of equation $\frac{d}{dt}(r \times v) = \frac{dr}{dt} \times v + r \times \frac{dv}{dt}$, dr/dt is simply the velocity v, leaving $v \times v$. Since the cross product of any vector with itself is always zero, that term drops out, leaving:

$$\frac{d}{dt}(r \times v) = \frac{dr}{dt} \times v + r \times \frac{dv}{dt}.$$

$$\frac{d}{dt}(r \times v) = r \times \frac{dv}{dt}.$$

Here, dv/dt is the acceleration a of the particle. Thus, if equation $\frac{d}{dt}(r \times v) = r \times \frac{dv}{dt}$ is multiplied by m, the left-hand side becomes dL/dt, as in equation $\frac{dL}{dt} = \frac{d}{dt}(r \times p) = m\frac{d}{dt}(r \times v)$, and the right-hand side may be written $r \times ma$. Since, according to Newton's second law, ma is equal to F, the net force acting on the particle, the result is:

$$\frac{a}{dt}(r \times v) = r \times \frac{av}{dt}.$$

$$\frac{aL}{dt} = \frac{a}{dt}(r \times p) = m\frac{a}{dt}(r \times v),$$

$$\frac{dL}{dt} = r \times F.$$

Equation $\frac{dL}{dt} = r \times F$ means that any change in the angular momentum of a particle must be produced by a force that is not acting along the same direction as r. One particularly important application is the solar system. Each planet is held in its orbit by its gravitational attraction to the Sun, a force that acts along the vector from the Sun to the planet. Thus, the force of gravity cannot change the angular momentum of any planet with respect to the Sun. Therefore, each planet has constant angular momentum with respect to the Sun. This conclusion is correct even though the real orbits of the planets are not circles but ellipses.

$$\frac{dL}{dt} = r \times F.$$

The quantity r × F is called the torque τ. Torque may be thought of as a kind of twisting force, the kind needed to tighten a bolt or to set a body into rotation. Using this definition, equation $\frac{dL}{dt} = r \times F$ may be rewritten:

$$\tau = r \times F = \frac{dL}{dt}.$$

Equation $\tau = r \times F = \frac{dL}{dt}$ means that if there is no torque acting on a particle, its angular momentum is constant, or conserved. Suppose, however, that some agent applies a force F_a to the particle resulting in a torque equal to $r \times F_a$. According to Newton's third law, the particle must apply a force $-F_a$ to the agent. Thus, there is a torque equal to $-r \times F_a$ acting on the agent. The torque on the particle causes its angular momentum to change at a rate given by $dL/dt = r \times F_a$. However, the angular momentum La of the agent is changing at the rate $dL_a/dt = -r \times F_a$. Therefore, $dL/dt + dL_a/dt = 0$, meaning that the total angular momentum of particle plus agent is constant, or conserved. This principle may be generalized to include all interactions between bodies of any kind, acting by way of forces of any kind. Total angular momentum is always conserved. The law of conservation of angular momentum is one of the most important principles in all of physics.

$$\tau = r \times F = \frac{dL}{dt}.$$

Motion of a Group of Particles

Centre of Mass

The word particle has been used in this article to signify an object whose entire mass is concentrated at a point in space. In the real world, however, there are no particles of this kind. All real bodies have sizes and shapes. Furthermore, as Newton believed and is now known, all bodies are in fact compounded of smaller bodies called atoms. Therefore, the science of mechanics must deal not only with particles but also with more complex bodies that may be thought of as collections of particles.

To take a specific example, the orbit of a planet around the Sun as if the planet and the Sun were each concentrated at a point in space. In reality, of course, each is a substantial body. However, because each is nearly spherical in shape, it turns out to be permissible, for the purposes of this problem, to treat each body as if its mass were concentrated at its centre. This is an example of an idea that is often useful in discussing bodies of all kinds: the centre of mass. The centre of mass of a uniform sphere is located at the centre of the sphere. For many purposes (such as the one cited above) the sphere may be treated as if all its mass were concentrated at its centre of mass.

To extend the idea farther, consider Earth and the Sun not as two separate bodies but as a single system of two bodies interacting with one another by means of the force of gravity. In the previous discussion of circular orbits, the Sun was assumed to be at rest at the centre of the orbit, but, according to Newton's third law, it must actually be accelerated by a force due to Earth that is equal and opposite to the force that the Sun exerts on Earth. In other words, considering only the Sun and Earth (ignoring, for example, all the other planets), if M_S and M_E are, respectively, the masses of the Sun and Earth, and if *aS* and *aE* are their respective accelerations, then combining Newton's second and third laws results in the equation $M_S a_S = -M_E a_E$. Writing each a as dv/dt, this equation is easily manipulated to give:

$$\frac{d}{dt}\left(M_S v_S + M_E v_E\right) = 0,$$

$$M_S v_S + M_E v_E = \text{constant}.$$

This remarkable result means that, as Earth orbits the Sun and the Sun moves in response to Earth's gravitational attraction, the entire two-body system has constant linear momentum, moving in a straight line at constant speed. Without any loss of generality, one can imagine observing the system from a frame of reference moving along with that same speed and direction. This is sometimes called the centre-of-mass frame. In this frame, the momentum of the two-body system—i.e., the constant in equation $M_S v_S + M_E v_E = \text{constant}$—is equal to zero. Writing each of the v's as the corresponding dr/dt, equation $M_S v_S + M_E v_E = \text{constant}$ may be expressed in the form:

$$M_S v_S + M_E v_E = \text{constant}.$$

$$\frac{d}{dt}\left(M_S r_S + M_E r_E\right) = 0.$$

Thus, $M_S r_S$ and $M_E r_E$ are two vectors whose vector sum does not change with time. The sum is defined to be the constant vector MR, where M is the total mass of the system and equals $M_S + M_E$. Thus,

$$MR = M_S r_S + M_E r_E$$

This procedure defines a constant vector *R*, from any arbitrarily chosen point in space. The relation between vectors *R*, r_S, and r_E is shown in figure. The fact that *R* is constant (although r_S and r_E are not constant) means that, rather than Earth orbiting the Sun, Earth and the Sun are both orbiting an imaginary point fixed in space. This point is known as the centre of mass of the two-body system.

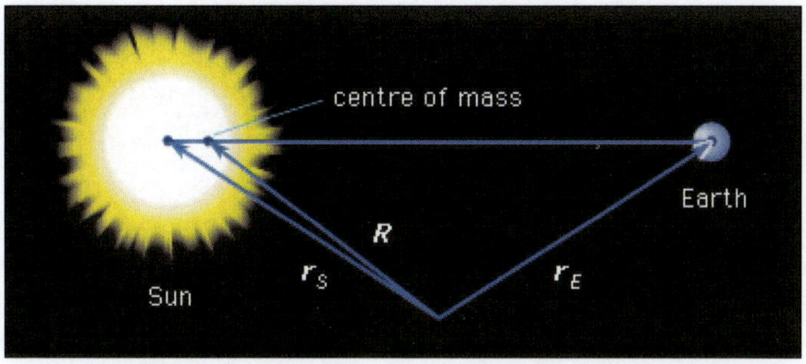

The centre of mass of the two-body Earth-Sun system.

Knowing the masses of the two bodies ($M_S = 1.99 \times 10^{30}$ kilograms, $M_E = 5.98 \times 10^{24}$ kilograms), it is easy to find the position of the centre of mass. The origin of the coordinate system may be chosen to be located at the centre of mass merely by defining $R = 0$. Then $r_S = (M_E / M_S) r_E \approx 450$ kilometres, when r_E is rounded to 1.5×10^8 km. A few hundred kilometres is so small compared to r_E that, for all practical purposes, no appreciable error occurs when r_S is ignored and the Sun is assumed to be stationary at the centre of the orbit.

With this example as a guide, it is now possible to define the centre of mass of any collection of bodies. Assume that there are N bodies altogether, each labeled with numbers ranging from 1 to N, and that the vector from an arbitrary origin to the ith body—where i is some number between 1 and N—is r_i, as shown in figure. Let the mass of the ith body be m_i. Then the total mass of the N-body system is:

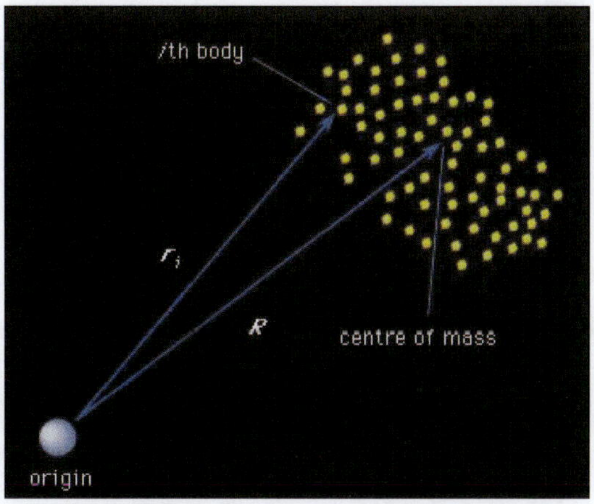

The centre of mass of an N-body system.

$$m = \sum_{i=1}^{N} m_i,$$

and the centre of mass of the system is found at the end of a vector *R* given by:

$$mR = \sum_{i=1}^{N} m_i r_i,$$

This definition applies regardless of whether the *N* bodies making up the system are the stars in a galaxy, the atoms in a rigid body, larger and arbitrarily chosen segments of a rigid body, or any other system of masses. According to equation $mR = \sum_{i=1}^{N} m_i r_i$, the vector to the centre of mass of any system is a kind of weighted average of the vectors to all the components of the system.

$$mR = \sum_{i=1}^{N} m_i r_i,$$

As will be demonstrated in the sections that follow, the statics and dynamics of many complicated bodies or systems may often be understood by simply applying Newton's laws as if the system's mass were concentrated at the centre of mass.

Conservation of Momentum

Newton's second law, in its most general form, says that the rate of a change of a particle's momentum p is given by the force acting on the particle; i.e., F = dp/dt. If there is no force acting on the particle, then, since dp/dt = 0, p must be constant, or conserved. This observation is merely a restatement of Newton's first law, the principle of inertia: if there is no force acting on a body, it moves at constant speed in a straight line.

Now suppose that an external agent applies a force F_a to the particle so that p changes according to:

$$\frac{dP}{dt} = F_a.$$

According to Newton's third law, the particle must apply an equal and opposite force $-F_a$ to the external agent. The momentum *pa* of the external agent therefore changes according to:

$$\frac{dP_a}{dt} = -F_a.$$

Adding together equations $\frac{dP}{dt} = F_a$ and $\frac{dP_a}{dt} = -F_a$ results in the equation:

$$\frac{dP}{dt} = F_a$$

$$\frac{dP_a}{dt} = -F_a.$$

$$\frac{d}{dt}(p + p_a) = 0.$$

The force applied by the external agent changes the momentum of the particle, but at the same time the momentum of the external agent must also change in such a way that the total momentum of both together is constant, or conserved. This idea may be generalized to give the law of conservation of momentum: in all the interactions between all the bodies in the universe, total momentum is always conserved.

It is useful in this light to examine the behaviour of a complicated system of many parts. The centre of mass of the system may be found using equation $mR = \sum_{i=1}^{N} m_i r_i$. Differentiating with respect to time gives:

$$mv = \sum_{i=1}^{N} m_i v_i,$$

where v = dR/dt and $v_i = dr_i / dt$. Note that $m_i v_i$ is the momentum of the ith part of the system, and mv is the momentum that the system would have if all its mass (i.e., m) were concentrated at its centre of mass, the point whose velocity is v. Thus, the momentum associated with the centre of mass is the sum of the momenta of the parts.

Suppose now that there is no external agent applying a force to the entire system. Then the only forces acting on the system are those exerted by the parts on one another. These forces may accelerate the individual parts. Differentiating equation $mv = \sum_{i=1}^{N} m_i v_i$, with respect to time gives:

$$mv = \sum_{i=1}^{N} m_i v_i,$$

$$m\frac{dv}{dt} = \sum_{i=1}^{N} m_i \frac{dv_i}{dt} = \sum_{i=1}^{N} F_i,$$

where F_i is the net force, or the sum of the forces, exerted by all the other parts of the body on the *i*th part. F_i is defined mathematically by the equation:

$$F_i = \sum_{j=1}^{N} F_{ij},$$

where F_{ij} represents the force on body *i* due to body *j* (the force on body *i* due to itself, F_{ii}, is zero). The motion of the centre of mass is then given by the complicated-looking formula:

$$m\frac{dv}{dt} = \sum_{i=1}^{N}\left(\sum_{j=1}^{N} F_{ij}\right).$$

This complicated formula may be greatly simplified, however, by noting that Newton's third law requires that for every force F_{ij} exerted by the *j*th body on the *i*th body, there is an equal and opposite force $-F_{ij}$ exerted by the ith body on the jth body. In other words, every term in the double sum has an equal and opposite term. The double summation on the right-hand side of equation $F_i = \sum_{j=1}^{N} F_{ij}$, always adds up to zero. This result is true regardless of the complexity of the system, the nature of the forces acting between the parts, or the motions of the parts. In short, in the absence of external forces acting on the system as a whole, $mdv/dt = 0$, which means that the momentum of the centre of mass of the system is always conserved. Having determined that momentum is conserved whether or not there is an external force acting, one may conclude that the total momentum of the universe is always conserved.

$$F_i = \sum_{j=1}^{N} F_{ij},$$

Collisions

A collision is an encounter between two bodies that alters at least one of their courses. Altering the course of a body requires that a force be applied to it. Thus, each body exerts a force on the other. These forces of interaction may operate at some distance, as do the gravitational and electromagnetic forces, or the bodies may appear to make physical contact. However, even apparent contact between two bodies is only a macroscopic manifestation of microscopic forces that act between atoms some distance apart. There is no fundamental distinction between physical contact and interaction at a distance.

The importance of understanding the mechanics of collisions is obvious to anyone who has ever driven an automobile. In modern physics, however, collisions are important for a different reason. The current understanding of the subatomic particles of which atoms are composed is derived entirely from studying the results of collisions among them. Thus, in modern physics, the description of collisions is a significant part of the understanding of matter. These descriptions are quantum mechanical rather than classical, but they are nevertheless closely based on principles that arise out of classical mechanics.

It is possible in principle to predict the result of a collision using Newton's second law directly. Suppose that two bodies are going to collide and that F, the force of interaction between them, is known to be a function of r, the distance between them. Then, if it is known that, say, one particle has incident momentum p, the problem is solved if the final momentum $p + \Delta p$ can be determined. Inverting Newton's second law, $F = dp/dt$, the change in momentum is given by:

$$\Delta p = \int_{-\infty}^{\infty} F dt.$$

This integral is known as the impulse imparted to the particle. In order to perform the integral, it is necessary to know r at all times so that F may be known at all times. More realistically, Δp is the sum of a series of small steps, such that:

$$\delta p = F \delta t,$$

where F depends on the instantaneous distance between the particles. Because $p = mv = mdr/dt$, the change in r in this step is:

$$\delta r = \frac{P}{m} \delta t.$$

At the next step, there is a new distance, $r + \delta r$, giving a new value of the force in equation and a new momentum, $p + \delta p$, in equation $\delta r = \frac{P}{m} \delta t$. This method of analyzing collisions is used in numerical calculations on digital computers.

$$\delta p = F \delta t,$$

$$\delta r = \frac{P}{m} \delta t$$

To predict the result of a collision analytically (rather than numerically) it is often most useful to apply conservation laws. In any collision (as in any other phenomenon), energy, momentum, and angular momentum are always conserved. Judicious application of these laws may be extremely useful because they do not depend in any way on the detailed nature of the interaction (i.e., the force as a function of distance).

This point can be illustrated by the following example. A collision is to take place between two bodies of the same mass m. One of the bodies is initially at rest (its momentum is zero). The other has initial momentum p_0. After the collision, the body previously at rest has momentum p_1, and the body initially in motion has momentum p_2. Since momentum is conserved, the total momentum after the collision, $p_1 + p_2$, must be equal to the total momentum before the collision, p_0; that is,

$$P_0 = P_1 + P_2$$

Equation $P_0 = P_1 + P_2$ is the equation of a vector triangle, as shown in figure. However, p_1 and p_2 are not determined by this condition; they are only constrained by it.

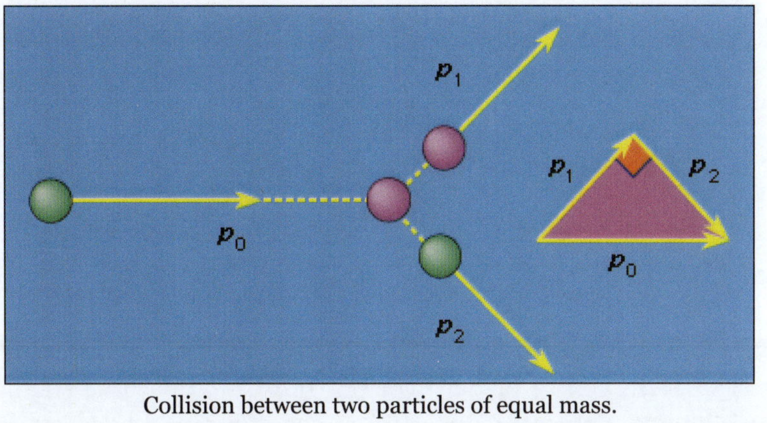

Collision between two particles of equal mass.

Although energy is always conserved, the kinetic energy of the incident body is not always converted entirely into the kinetic energy of the two bodies after the collision. For example, if the bodies are microscopic (say, two identical atoms), the collision may cause one or both to be excited into a state of higher internal energy than it started with. Such an event would leave correspondingly less kinetic energy for the outgoing atoms. In fact, it is precisely by studying the trajectories of outgoing projectiles in collisions like these that physicists are able to determine the possible excited states of microscopic particles.

In a collision between macroscopic objects, some of the kinetic energy is always converted to heat. Heat is the energy of random vibrations of the atoms and molecules that constitute the bodies. However, if the amount of heat is negligible compared to the initial kinetic energy, it may be ignored. Such a collision is said to be elastic.

Suppose the collision described above between two bodies, each of mass m, is between billiard balls, and suppose it is elastic (a reasonably good approximation of real billiard balls). The kinetic energy of the incident ball is then equal to the sum of the kinetic energies of the outgoing balls. According to equation $K = \frac{1}{2}mv^2$, the kinetic energy of a moving object is given by $K = {}^1/_2 mv^2$, where v is the speed of the ball. Equation $K = \frac{1}{2}mv^2$ may be written in a particularly useful form by recognizing that since $p = mv$,

$$K = \frac{1}{2}mv^2$$

$$K = \frac{1}{2}mv^2 = \frac{p^2}{2m}.$$

Then the conservation of kinetic energy may be written:

$$\frac{P_0^2}{2m} = \frac{P_1^2}{2m} + \frac{P_2^2}{2m},$$

or, canceling the factors 2m,

$$P_0^2 = P_1^2 + P_2^2.$$

Comparing this result with equation $P_0 = P_1 + P_2$ shows that the vector triangle is pythagorean; p_1 and p_2 are perpendicular. This result is well known to all experienced pool players. Notice that it was possible to arrive at this result without any knowledge of the forces that act when billiard balls collide.

Relative Motion

A collision between two bodies can always be described in a frame of reference in which the total momentum is zero. This is the centre-of-mass (or centre-of-momentum) Then, for example, in the collision between two bodies of the same mass discussed above, the two bodies always have equal and opposite velocities, as shown in figure. It should be noted that, in this frame of reference, the outgoing momenta are antiparallel and not perpendicular.

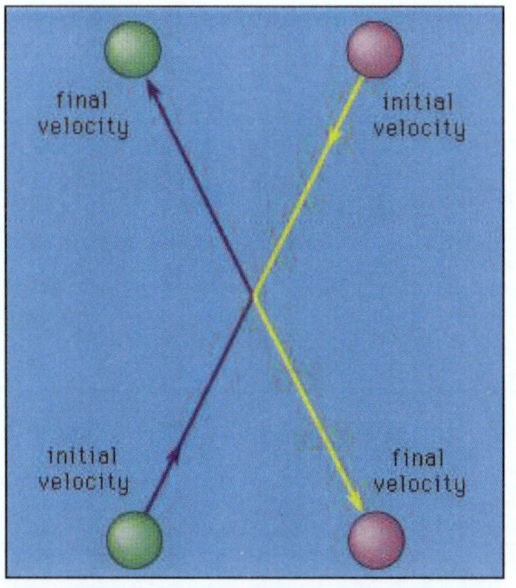

Collision between two particles of equal mass as seen from the centre-of-mass frame of reference.

Any collection of bodies may similarly be described in a frame of reference in which the total momentum is zero. This frame is simply the one in which the centre of mass is at rest. This fact is easily seen by differentiating equation $mR = \sum_{i=1}^{N} m_i r_i$, with respect to time, giving:

$$m\frac{dR}{dt} = \sum_{i=1}^{N} m_i \frac{d\,r_i}{dt}.$$

The right-hand side is the sum of the momenta of all the bodies. It is equal to zero if the velocity of the centre of mass, dR/dt, is equal to zero.

If Newton's second law is correct in any frame of reference, it will also appear to be correct to an observer moving with any constant velocity with respect to that frame. This principle, called the principle of Galilean relativity, is true because, to the moving observer, the same constant velocity seems to have been added to the velocity of every particle in the system. This change does not affect the accelerations of the particles (since the added velocity is constant, not accelerated) and therefore does not change the apparent force (mass times acceleration) acting on each particle. That is why it is permissible to describe a problem from the centre-of-momentum frame (provided that the centre of mass is not accelerated) or from any other frame moving at constant velocity with respect to it.

If this principle is strictly correct, the fundamental forces of physics should not contain any particular speed. This must be true because the speed of any object will be different to observers in different but equally good frames of reference, but the force should always be the same. It turns out, according to the theory of James Clerk Maxwell, that there is an intrinsic speed in the force laws of electricity and magnetism: the speed of light appears in the forces between electric charges and between magnetic poles. This discrepancy was ultimately resolved by Albert Einstein's special theory of relativity. According to the special theory of relativity, Newtonian mechanics breaks down when the relative speed between particles approaches the speed of light.

Coupled Oscillators

The motion of a group of particles bound by springs to one another is discussed. The solutions of this seemingly academic problem have far-reaching implications in many fields of physics. For example, a system of particles held together by springs turns out to be a useful model of the behaviour of atoms mutually bound in a crystalline solid.

To begin with a simple case, consider two particles in a line, as shown in figure. Each particle has mass m, each spring has spring constant k, and motion is restricted to the horizontal, or x, direction. Even this elementary system is capable of surprising behaviour, however. For instance, if one particle is held in place while the other is displaced, and then both are released, the displaced particle immediately begins to execute simple harmonic motion. This motion, by stretching the spring between the particles, starts to excite the second particle into motion. Gradually the energy of motion passes from the first particle to the second until a point is reached at which the first particle is at rest and only the second is oscillating. Then the process starts all over again, the energy passing in the opposite direction.

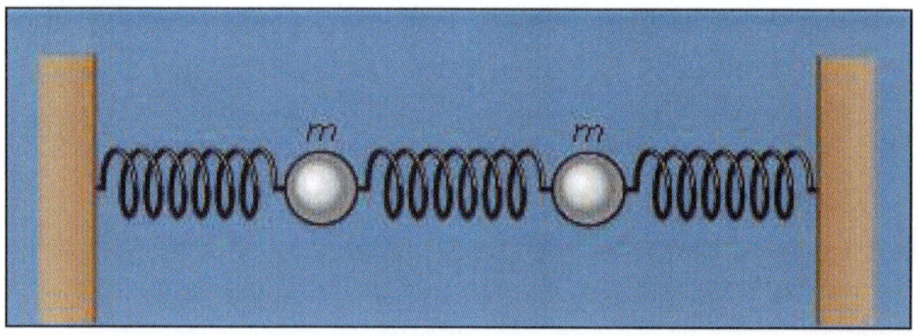

Coupled oscillators.

To analyze the possible motions of the system, one writes equations similar to equation $a = \frac{d^2x}{dt^2} = -\frac{k}{m}x$, giving the acceleration of each particle owing to the forces acting on it. There is one equation for each particle (two equations in this case). The force on each particle depends not only on its displacement from its equilibrium position but also on its distance from the other particle, since the spring between them stretches or compresses according to that distance. For this reason the motions are coupled, the solution of each equation (the motion of each particle) depending on the solution of the other (the motion of the other).

Analyzing the system yields the fact that there are two special states of motion in which both particles are always in oscillation with the same frequency. In one state, the two particles oscillate in opposite directions with equal and opposite displacements from equilibrium at all times. In the other state, both particles move together, so that the spring between them is never stretched or compressed. The first of these motions has higher frequency than the second because the centre spring contributes an increase in the restoring force.

These two collective motions, at different, definite frequencies, are known as the normal modes of the system.

If a third particle is inserted into the system together with another spring, there will be three equations to solve, and the result will be three normal modes. A large number N of particles in a line will have N normal modes. Each normal mode has a definite frequency at which all the particles oscillate. In the highest frequency mode each particle moves in the direction opposite to both of its neighbours. In the lowest frequency mode, neighbours move almost together, barely disturbing the springs between them. Starting from one end, the amplitude of the motion gradually builds up, each particle moving a bit more than the one before, reaching a maximum at the centre, and then decreasing again. A plot of the amplitudes, shown in figure, basically describes one-half of a sine wave from one end of the system to the other. The next mode is a full sine wave, then 3/2 of a sine wave, and so on to the highest frequency mode, which may be visualized as N/2 sine waves. If the vibrations were up and down rather than side to side, these modes would be identical to the fundamental and harmonic vibrations excited by plucking a guitar string.

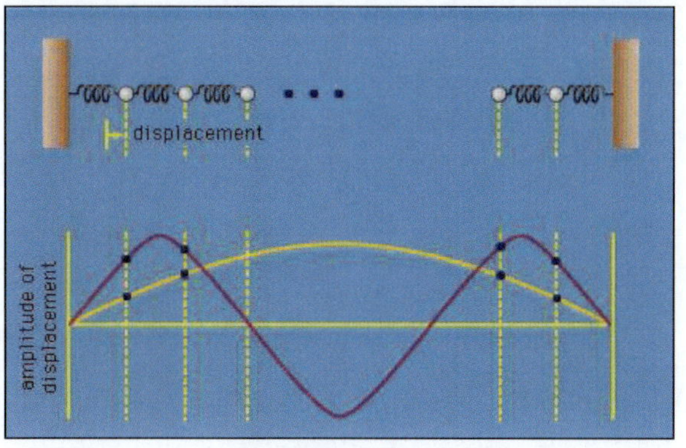

Normal modes.

The atoms of a crystal are held in place by mutual forces of interaction that oppose any disturbance from equilibrium positions, just as the spring forces in the example above. For small displacements of the atoms, they behave mathematically just like spring forces—i.e., they obey Hooke's law, equation $F = -kx$. Each atom is free to move in three dimensions rather than one, however; therefore each atom added to a crystal adds three normal modes. In a typical crystal at ordinary temperature, all these modes are always excited by random thermal energy. The lower-frequency, longer-wavelength modes may also be excited mechanically. These are called sound waves.

Analytic Approaches

Classical mechanics can, in essence, be reduced to Newton's laws, starting with the second law, in the form:

$$F = \frac{dp}{dt}.$$

If the net force acting on a particle is *F*, knowledge of *F* permits the momentum *p* to be found; and knowledge of *p* permits the position *r* to be found, by solving the equation:

$$\frac{dr}{dt} = \frac{P}{m}.$$

These solutions give the components of p—that is, p_x, p_y, and p_z —and the components of r—x, y, and z—each as a function of time. To complete the solution, the value of each quantity— p_x, p_y, p_z, x, y, and z—must be known at some definite time, say, t = 0. If there is more than one particle, an equation in the form of equation $F = \frac{dp}{dt}$ must be written for each particle, and the solution will involve finding the six variables x, y, z, p_x, p_y, and p_z for each particle as a function of time, each once again subject to some initial condition. The equations may not be independent, however. For example, if the particles interact with one another, the forces will be related by Newton's third law. In this case (and others), the forces may also depend on time.

If the problem involves more than a very few particles, this method of solution quickly becomes intractable. Furthermore, in many cases it is not useful to express the problem purely in terms of particles and forces. Consider, for example, the problem of a sphere or cylinder rolling without slipping on a plane surface. Rolling without slipping is produced by friction due to forces acting between atoms in the rolling body and atoms in the plane, but the interactions are very complex; they probably are not fully understood even today, and one would like to be able to formulate and solve the problem without introducing them or needing to understand them. For all these reasons, methods that go beyond solving equations $F = \frac{dp}{dt}$ and $\frac{dr}{dt} = \frac{P}{m}$ have had to be introduced into classical mechanics.

The methods that have been introduced do not involve new physics. In fact, they are deduced directly from Newton's laws. They do, however, involve new concepts, new language to describe those concepts, and the adoption of powerful mathematical techniques. Some of those methods are briefly surveyed here.

Configuration Space

The position of a single particle is specified by giving its three coordinates, *x*, *y*, and *z*. To specify the positions of two particles, six coordinates are needed, $x_1, y_1, z_1, x_2, y_2, z_2$. If there are *N* particles, 3*N* coordinates will be needed. Imagine a system of 3*N* mutually orthogonal coordinates in a 3*N*-dimensional space (a space of more than three dimensions is a purely mathematical construction, sometimes known as a hyperspace). To specify the exact position of one single point in this space, 3*N* coordinates are needed. However, one single point can represent the entire configuration of all *N* particles in the problem. Furthermore, the path of that single point as a function of time is the complete solution of the problem. This 3*N*-dimensional space is called configuration space.

Configuration space is particularly useful for describing what is known as constraints on a problem. Constraints are generally ways of describing the effects of forces that are best not explicitly introduced into the problem. For example, consider the simple case of a falling body near the surface of the Earth. The equations of motion—equations $z = z_0 - \frac{1}{2}gt^2$, $v = gt$, and $a = g$—are valid only until the body hits the ground. Physically, this restriction is due to forces between atoms in the falling body and atoms in the ground, but, as a practical matter, it is preferable to say that the solutions are valid only for z > 0 (where z = 0 is ground level). This constraint, in the form of an inequality, is very difficult to incorporate directly into the equations of the problem. In the language of configuration space, however, one merely needs to specify that the problem is being solved only in the region of configuration space for which z > 0.

Rolling without sliding on a plane, cannot easily be described in configuration space, since it is basically a condition on relative velocities of rotation and translation; but another constraint, that the body is restricted to motion along the plane, is easily described in configuration space.

Another type of constraint specifies that a body is rigid. Then, even though the body is composed of a very large number of atoms, it is not necessary to find separately the *x*, *y*, and *z* coordinate of each atom because these are related to those of the other atoms by the condition of rigidity. A careful analysis yields that, rather than needing 3*N* coordinates (where *N* may be, for example, 10^{24} atoms), only 6 are needed: 3 to specify the position of the centre of mass and 3 to give the orientation of the body. Thus, in this case, the constraint has reduced the number of independent coordinates from 3*N* to 6. Rather than restricting the behaviour of the system to a portion of the original 3*N*-dimensional configuration space, it is possible to describe the system in a much simpler 6-dimensional configuration space. It should be noted, however, that the six coordinates are not necessarily all distances. In fact, the most convenient coordinates are three distances (the *x, y*, and *z* coordinates of the centre of mass of the body) and three angles, which specify the orientation of a set of axes fixed in the body relative to a set of axes fixed in space. This is an example of the use of constraints to reduce the number of dynamic variables in a problem (the *x, y*, and *z* coordinates of each particle) to a smaller number of generalized dynamic variables, which need not even have the same dimensions as the original ones.

The Principle of Virtual Work

A special class of problems in mechanics involves systems in equilibrium. The problem is to find the configuration of the system, subject to whatever constraints there may be, when all forces are balanced. The body or system will be at rest (in the inertial rest frame of its centre of mass), meaning that it occupies one point in configuration space for all time. The problem is to find that point. One criterion for finding that point, which makes use of the calculus of variations, is called the principle of virtual work.

According to the principle of virtual work, any infinitesimal virtual displacement in configuration space, consistent with the constraints, requires no work. A virtual displacement means an instantaneous change in coordinates (a real displacement would require finite time during which particles might move and forces might change). To express the principle, label the generalized coordinates $r_1, r_2, \ldots, r_i, \ldots$ Then if Fi is the net component of generalized force acting along the coordinate r_i,

$$\sum_i F_i dr_i = 0.$$

Here, F_i dr_i is the work done when the generalized coordinate is changed by the infinitesimal amount dr_i. If r_i is a real coordinate (say, the x coordinate of a particle), then F_i is a real force. If r_i is a generalized coordinate (say, an angular displacement of a rigid body), then F_i is the generalized force such that F_i dr_i is the work done (for an angular displacement, F_i is a component of torque).

Take two simple examples to illustrate the principle. First consider two particles that are restricted to motion in the *x* direction and are constrained by a taut string connecting them. If their *x* coordinates are called x_1 and x_2, then $F_1 dx_1 + F_2 dx_2 = 0$ according to the principle of virtual work. But the taut string requires that the particles be displaced the same amount, so that $dx_1 = dx_2$, with the result that $F_1 + F_2 = 0$. The particles might be in equilibrium, for example, under equal and opposite forces, but F_1 and F_2 do not need individually to be zero. This is generally true of the *Fi* in equation $\sum_i F_i dr_i = 0$. As a second example, consider a rigid body in space. Here, the constraint of rigidity has already been expressed by reducing the coordinate space to that of six generalized coordinates.

These six coordinates (*x, y, z,* and three angles) can change quite independently of one another. In other words, in equation $\sum_i F_i dr_i = 0$, the six dr_i are arbitrary. Thus, the only way equation $\sum_i F_i dr_i = 0$ can be satisfied is if all six F_i are zero. This means that the rigid body can have no net component of force and no net component of torque acting on it. Of course, this same conclusion was reached earlier by less abstract arguments.

$$\sum_i F_i dr_i = 0.$$

Lagrange's and Hamilton's Equations

Elegant and powerful methods have also been devised for solving dynamic problems with constraints. One of the best known is called Lagrange's equations. The Lagrangian L is defined as $L = T - V$, where T is the kinetic energy and V the potential energy of the system in question. Generally speaking, the potential energy of a system depends on the coordinates of all its particles; this may be written as $V = V(x_1, y_1, z_1, x_2, y_2, z_2, \ldots)$. The kinetic energy generally depends on the velocities, which, using the notation $vx = dx/dt = \dot{x}$, may be written $T = T(\dot{x}_1, \dot{y}_1, \dot{z}_1, \dot{x}_2, \dot{y}_2, \dot{z}_2, \ldots)$. Thus, a dynamic problem has six dynamic variables for each particle—that is, *x, y, z* and $\dot{x}$, $\dot{y}$, $\dot{z}$—and the Lagrangian depends on all 6*N* variables if there are *N* particles.

In many problems, however, the constraints of the problem permit equations to be written relating at least some of these variables. In these cases, the 6*N* related dynamic variables may be reduced to a smaller number of independent generalized coordinates (written symbolically as $q_1, q_2, \ldots q_i, \ldots$) and generalized velocities (written as $\dot{q}_1, \dot{q}_2, \ldots \dot{q}_i, \ldots$), just as, for the rigid body, 3*N* coordinates were reduced to six independent generalized coordinates (each of which has an associated velocity). The Lagrangian, then, may be expressed as a function of all the q_i and $\dot{q}_i$. It is possible, starting from Newton's laws only, to derive Lagrange's equations:

$$\frac{d}{dt}\frac{\partial L}{\partial \dot{q}_i} - \frac{\partial L}{\partial q_i} = 0,$$

where the notation $\partial L/\partial q_i$ means differentiate L with respect to q_i only, holding all other variables constant. There is one equation of the form for each of the generalized coordinates q_i (e.g., six equations for a rigid body), and their solutions yield the complete dynamics of the system. The use of generalized coordinates allows many coupled equations of the form to be reduced to fewer, independent equations of the form.

$$\frac{d}{dt}\frac{\partial L}{\partial \dot{q}_i} - \frac{\partial L}{\partial q_i} = 0,$$

There is an even more powerful method called Hamilton's equations. It begins by defining a generalized momentum p_i, which is related to the Lagrangian and the generalized velocity $\dot{q}_i$ by $p_i = \partial L / \partial \dot{q}_i$. A new function, the Hamiltonian, is then defined by $H = \Sigma i\, \dot{q}_i p_i - L$. From this point it is not difficult to derive:

$$\dot{q}_i = \frac{\partial H}{\partial p_i}$$

And

$$-\dot{p}_i = \frac{\partial H}{\partial q_i}.$$

These are called Hamilton's equations. There are two of them for each generalized coordinate. They may be used in place of Lagrange's equations, with the advantage that only first derivatives—not second derivatives—are involved.

The Hamiltonian method is particularly important because of its utility in formulating quantum mechanics. However, it is also significant in classical mechanics. If the constraints in the problem do not depend explicitly on time, then it may be shown that H = T + V, where T is the kinetic energy and V is the potential energy of the system—i.e., the Hamiltonian is equal to the total energy of the system. Furthermore, if the problem is isotropic (H does not depend on direction in space) and homogeneous (H does not change with uniform translation in space), then Hamilton's equations immediately yield the laws of conservation of angular momentum and linear momentum, respectively.

Mathematical Formulation of Newtonian Mechanics

For the dynamics of a body small enough (relative to an entire system) to be treated as a single particle, this principle can be expressed mathematically as an ordinary differential equation with time as the independent variable. Motions of planets and of their satellites, for example, lend themselves to interpretation in terms of solutions of such ordinary differential equations.

In other circumstances, when matter is viewed as subdivided into many particles, even the laws of statics may need to be expressed as partial differential equations with space coordinates as independent variables. Their form depends upon an assumed knowledge of material properties, including knowledge of how internal stresses (the internal forces acting between adjacent particles, divided by their area of contact) depend upon other factors including the body's degree of deformation. Such a partial differential equation may express the fact that the sum of all forces on a particle, including external forces as well as the internal forces of interaction with adjacent particles, is zero.

Dynamics, on the other hand, allows that sum to take a nonzero value which determines how the momentum of the particle concerned changes from one brief instant to the next. Analytical expression of results such as this takes the form of partial differential equations with space coordinates and also the time as independent variables. Modern treatments of such problems using high-speed computers may approximate the partial differential equations by means of finite differences; or may, alternatively, apply the basic principles of mechanics directly to carefully selected particles of matter known as finite elements.

Theoretical mechanics has continued over the years to make massive progress through (i) improved representation of material properties, (ii) incorporation of additional types of external force, and (iii) advances in analytical or numerical methods. Brief indications of these main lines of progress will now be given.

Improved Representation of Material Properties

For matter in the solid state a rather early initiation, and subsequent extended development, of a comprehensive theory of elasticity, allowing stress to depend linearly on the measure of deformation known as strain, led to impressive results in both statics and dynamics. It was then followed successively by a range of more refined theories (including, for example, the theory of plasticity) which especially need to be applied in cases of large deformation. Important newer developments have resulted from the treatment of 'materials with memory'.

For matter in either of the two fluid states (liquid or gas), no serious problems are posed by statics; mainly, because stresses in a fluid at rest take the form of a simple pressure acting equally in all directions, as explained in elementary hydrostatics texts. The dynamics of fluids, however, is affected by additional stresses associated for common fluids with their viscosity: and such stresses prove to be important even for fluids of very small viscosity. This is because the condition satisfied by a fluid motion at a solid boundary requires its velocity to vary steeply across a thin boundary layer, where viscous stresses are substantial and may significantly affect whether or not the boundary layer, as a result of separating prematurely from such a solid boundary, changes drastically the character of the whole flow.

Different relationships of stress to motion, more complicated than can be described by a simple viscosity coefficient, govern the flow properties of certain substances. Analysis of these relationships and their consequences forms the branch of mechanics known as rheology.

Other material properties which may be important include those familiar from thermodynamics. They are concerned with the convertibility of energy between its various forms, and with 'equations of state'. In a fluid, for example, these relate changes in the fluid density (mass per unit volume) to changes in pressure and also temperature. For gases, furthermore, kinetic theory (applying mechanical laws to the molecules themselves) may give valuable information about material properties which can critically affect gas dynamics, including not only viscous action but also heat transfer (which, besides influencing dynamical behaviour, is also of importance in its own right).

Consideration of Additional External Forces

External forces which in mechanics may need to be considered together with gravity include electromagnetic forces; whose effect can be substantial, especially for good conductors. These include both solid and liquid metals, and also ionized gases, some of whose electromagnetic properties are found to deserve study by means of kinetic theory.

Systems subjected to a general rotation may be analyzed by means of space coordinates related to a rotating frame of reference. Dynamics in such a rotating frame effectively incorporates additional forces (the centrifugal and Coriolis forces) which influence greatly the behaviour of technically important gyroscopic systems, and also the dynamics of the ocean or the atmosphere.

Advances in Mathematical Methods

Beginning with the analytical mechanics of Lagrange, mathematical theories founded upon energy considerations have proved to be of very general and wide-spread applicability. Some special mathematical methods of importance to theoretical mechanics include linear and nonlinear theories of vibrations and waves. These, for example, have proved to be of great value for acoustics by elucidating not only the types of vibrations which our ears detect but also their transmission to our ears as waves travelling through solid or fluid media.

Theories of stability are also most important. For statics, they can distinguish between an equilibrium which can persist and one which will collapse in response to a small disturbance of a particular shape (as when a solid shell buckles under load). For dynamics, they can distinguish between steady motions able to persist and those which will break up through instability into either regular or randomised disturbances. Description of instability, phenomena in terms of the generic process known as 'bifurcation' have proved increasingly fruitful.

More recently, general studies of dynamical systems have given the broad analytical background to two important tendencies: the tendency for certain continuously varying motions to develop discontinuities ('catastrophes', as when an acoustic wave develops into a shock wave); and the tendency for certain well ordered motions to develop into random motions ('chaos'; a process long studied in the dynamics of fluids as the tendency for laminar flows to develop into turbulence, but now recognized as relevant also to many solid systems with only a few degrees of freedom). Much of the previously existing knowledge and understanding in these areas (for example, of shock waves or turbulence) has been valuably extended and systematized by these new theories.

In all periods, many important mathematical advances have been initiated by research workers in mechanics. Thus the science of mathematics itself was a major beneficiary from progress in theoretical mechanics.

Examples of the Applications of Mechanics

Applied Mechanics and its Growing Utilisation of Theoretical Mechanics

The applications of mechanics are found in many scientific fields, some of which (like astronomy, oceanography, and meteorology) have already been referred to; as well as in most of the principal subdivisions of engineering and technology. Into each particular scientific or technological field the process of penetration of any refined ideas from theoretical mechanics has been slow, for a good reason: the scientific phenomena needing to be elucidated, or the technological objectives needing to be met, were in most cases much too complicated for effective treatment by the methods of theoretical mechanics during their early stages of development.

In these circumstances, practitioners of the disciplines concerned needed to concentrate above all on devising ingenious measurement techniques appropriate to a particular area of application, and on amassing useful empirical rules for correlating data obtained with these techniques. Such rules might be expressed in phraseology using some of the simpler ideas from theoretical mechanics, but they were not based on detailed theoretical analysis.

Great achievements were to result from applied mechanics development using these empirical methods. In each area the approach in question has, furthermore, continued to make good progress, especially during periods immediately following the introduction of a new measurement technique.

At the same time the general progress in theoretical mechanics has successfully taken place, partly under its own momentum and partly under the stimulus of challenges posed by the complicated problems needing to be solved in particular areas of application. Gradually, within each field, theoretical analysis of 'model problems' has become recognized as making a truly valuable contribution to studies of the complicated 'real problems' which they were designed to model. Even when the agreement between experiment and theory was not very excellent such analysis might permit a most useful extrapolation of the available experimental data to other conditions for which additional experiments would be too difficult or costly or time-consuming. In time, the models became more and more comprehensive, and correspondingly more valuable.

These continuing processes, in which theoretical mechanics both makes valued contributions to, and is stimulated by challenges derived from, applied mechanics, have increasingly brought the two parts of mechanics much closer together in a fruitfully cooperative unity. Some examples of how this happened in particular areas of application are given in the rest of this note.

Structural Engineering

In structural engineering, early empirical developments had led to great achievements which included such refined designs as medieval cathedrals. Yet gradually methods became more analytical as it became possible to estimate loads and the stress distributions they would produce, first in single beams and then in frameworks; and to compare stresses so calculated with yield stresses characteristic of the materials used. Stability calculations were initiated too, by Euler, so as to yield estimates of critical loads for buckling.

Nevertheless, over long periods it was necessary to overdesign structures by 'factors of safety' as high as 10; that is, to design them so that calculated loads for yielding or buckling were ten times more than the estimated worst-case loads. This need arose from a great multiplicity of uncertainties; for example, in the field of stress analysis, or concerning the reliability of assumptions about material properties or load estimates.

Later, the study of frameworks and similar structures became much more thorough; it took increasingly into account the problems of stress concentrations at junctions between elements, and of detailed design to minimise these. Also, 'limit design' methods were developed (studying the behaviour of a framework after early applications of load have produced certain plastic deformations).

Furthermore, the inevitable presence in a structure of particular types of imperfection which adversely influence structural integrity was increasingly allowed for. These included geometrical imperfections powerfully affect buckling behaviour.

Again, small material imperfections including dislocations and cracks were intensely studied in relation to their capacity for growth leading to plastic deformation or to fracture. This, for example, allowed the phenomenon of metal fatigue under cyclic loading (first discovered empirically) to become clearly interpretable in terms of crack growth, and there were similar successes in the area of metal creep. In the meantime, the importance of composite materials (from concrete to carbon-fiber-reinforced plastics) became increasingly recognized, and this too stimulated many major new developments in theoretical mechanics.

The enormous economic advantage of structure-weight minimisation for the design of aircraft and spacecraft gave a special boost to refined structural analysis aimed at achieving accuracies such as would permit 'factors of safety' to be brought down from former figures like 10 to modern figures around 1.5 or even less. They also encouraged particular mathematical developments (for example, in the theory of shells) of very great subtlety.

At the same time, empirical methods advanced in parallel; particularly through the development of strain-gauge technology, of various optical techniques and of a wide range of convenient methods for loading simulation. The modern structural engineer uses an admirably well integrated blend of empirical activity with the analytic and computational methods of theoretical mechanics.

Hydraulics

The study of flow in man-made systems of pipes and channels goes back to the beginnings of history with the irrigation achievements in Egypt and Mesopotamia which made possible the first great civilisations.

Theoretical mechanics made its impact on the subject in the eighteenth century when Daniel Bernoulli showed how a steady stream, according to Newton's Laws, would in the absence of frictional forces conserve what we now call its total head. Much of the ensuing development of hydraulics (at first, mainly empirical) was expressed in terms of laws governing how much total head is frictionally dissipated by different types of steady stream under different conditions. Some puzzling features of these laws, which recognize (for example) how streams progressing along pipes lose total head far less steeply where the pipe cross-section contracts than where it expands, became intelligible in terms of theoretical mechanics only when the conditions governing boundary layer separation in a fluid flow had been elucidated both theoretically and empirically.

Other developments in the two-dimensional (and three-dimensional) modelling of fluid motions allowed improved design of pipe and channel junctions and detailed calculations of flows such as those in the neighbourhood of sluice-gates. More recently, turbulence modelling played a growing part in such work. In the meantime, the increasingly refined theory of surface wave phenomena had been most valuably applied so as to understand how stationary waves, standing waves and travelling waves are generated by flows in open channels. Also, cavitation theory elucidated effects due to the motion of a liquid generating locally negative pressures which cause the spontaneous appearance of bubbles.

Mechanical Engineering

Building upon the long-established empirical technology of water-powered and wind-powered mill machinery, mechanical engineering began to develop rapidly after steam-power was introduced in the eighteenth century. Ideas from theoretical mechanics played an increasing role in these developments.

Thus, the part of dynamics known as kinematics (the geometrical description of relative motions) influenced the detailed design of trains of gearing mechanisms of all kinds; and, especially, the shaping of surfaces required to be in contact from gear-teeth to screws. The part known as kinetics (relation of motion to forces and to the work done by these) influenced the design not only of the actual means for transferring power but also of the devices for controlling the resulting motions. An increasingly important contribution was made by the theory of vibrations of mechanical systems, with its fundamental elucidation of the concept of normal modes of vibration each with its own natural frequency; and of the dominating influence of resonance (coincidence between forcing frequency and natural frequency) in generating these usually unwanted vibrations.

By the last quarter of the nineteenth century the mechanics of fluids had begun to exert a major influence on mechanical engineering design; as when, for example, a hydrodynamic theory of lubrication related the forces sustainable in a journal bearing to the geometry and viscous properties of the oil film. Above all, development of the steam turbine called for meticulous shaping of steam passages and of turbine blades, taking into account the interacting dynamics and thermodynamics of fast-moving steam. A new understanding of heat transfer, based on the fundamental mechanics of fluids, was at the same time being established.

Systems for efficient exchange of energy (in either direction) between streams of fluid and rotating blades were then progressively developed for the successive requirements of fans, propellers, fluid transmissions, compressors and gas turbines; with applications in aeronautics and many other fields. More and more, these developments rested upon refined representations of the dynamics of fluids and blades in relative motion. Other flow geometries needed to be analysed as the great liquid-fuelled rocket engines began to be designed. In many such developments the fluid dynamics of combustion played a critical part; while, once again, the associated cooling systems needed careful study by fluid-dynamic heat transfer theory. The same ingredients (chemical reactions, heat transfer and fluid dynamics, with the dynamics of mixing playing a dominant role) underlie many current developments in process engineering.

In the meantime the modern kinematic and kinetic analysis of machines and mechanisms has continued to make great progress, which has proved especially vital for the increasingly important technology of robotics. Gyro dynamics, again, has led to remarkable developments in control and navigation through the design of sensitive systems combining gyroscopes and accelerometers.

The modern mechanical engineer, to be sure, synthesizes knowledge from many different disciplines; including (for example) an even wider area of materials science than was touched on earlier in relation to structural engineering, and including also large parts of electronics that are essential for purposes both of instrumentation and control. Theoretical mechanics of solids and fluids continues, however, to form the fundamental basis of the mechanical engineer's art.

External Fluid Dynamics

The interaction of a solid body with an external fluid through which it moves is the subject of external fluid dynamics. Remarkable developments in external aerodynamics (the case where the fluid is air) were needed, in addition to several structural and power-plant developments that have been referred to earlier, to make possible the twentieth century's extraordinary achievements in aeronautical engineering.

Efficient horizontal flight requires the air pressures and viscous stresses around an aircraft to produce a resultant force with a large vertical component or lift to balance the weight, and yet with a low horizontal component or drag to be overcome by engine thrust. The viscosity of air is small but in external aerodynamics must not be neglected since classical theories on this assumption predict a zero resultant force on any solid body in steady motion through air. Accordingly, the discoveries by Prandtl and others, that well designed external shapes for aircraft could allow all the effects of viscosity to be confined to thin boundary layers and wakes, permitting drag to be low although (when the wakes contained intense trailing vortices) lift could be high, represented a vital first step forward in aerodynamic design; allowing major improvements in performance over that of the crude shapes of aviation's early years.

Later, after improvements in engines and in structural design as well as in aero-dynamics had raised aircraft speeds near to or above the speed of sound, aircraft began to experience an additional component of drag, associated with generation of an acoustic wave often called the 'sonic boom'. The corresponding feature in external hydrodynamics is observed when a ship's speed increases so that it experiences not only a wake-generating drag but also a wave-making drag. Wave theory for both acoustic waves in air and surface waves on water has been fruitfully applied to the study of shapes that will postponed, until substantially increased speeds, any significant rise in drag due to wave-making.

Other important parts of external fluid dynamics are concerned with the stability and control of the motion of aircraft through air or of ships or submarines through water. These parts achieved great success through applying the laws of dynamics to the solid vehicle in each case, taking into account fluid forces as affected by the vehicle's movement and by the action of control surfaces such as rudders, ailerons, etc.

At very high speeds, the integrity of an aircraft's structure may be threatened by frictional generation of heat in the boundary layer and this may once more demand careful heat transfer analyses. In the case of spacecraft re-entering the earth's atmosphere the problems are intensified and the analyses need to be at a very refined level.

Ingenious pressure-measuring and flow-visualisation techniques using copiously instrumented wind tunnels and other experimental facilities have been essential to the successes achieved in external fluid dynamics. Recently, the complete probing of a complicated velocity field in a fluid became possible through the development of Laser Doppler anemometry. Yet simultaneous improvements in computational fluid dynamics, based on major advances in fluid dynamic theory, have taken such a form that the balance of a designer's reliance on the twin supports of experimental data and theoretical analysis has continued to shift slowly but progressively in the direction of theory.

Planetary Sciences

After the astronomers of the sixteenth and early seventeenth centuries had amassed a remarkable body of accurate planetary observations, and Kepler had established a post-Copernican interpretation of these in terms of three empirical laws satisfied by the orbits of planets (and of their satellites), a first great success of the theoretical mechanics introduced by Newton was his demonstration that those laws must be interpreted in terms of an inverse-square law of gravitational attractive force. Much later, simultaneous attraction by more than one body was approximately allowed for, and used by Euler, Lagrange, Laplace and others to explain the principal departures from Kepler's laws which more accurate observations had brought to light.

The perturbation theory developed for this purpose proved invaluable after artificial satellites of earth were launched from 1959 onwards. Careful observation of the slow deviation of their orbits from those predicted by Kepler's laws allowed accurate determinations of the perturbing forces on a satellite (that is, forces other than simple attraction to the earth's centre). Important geophysical information was so obtained, including the best available data on departures of earth's shape from the spherical, and on winds in those highest parts of the atmosphere which include the orbits of many satellites.

Study of the lower parts of the earth's atmosphere is, of course, the province of meteorology. The invention of the barometer in the seventeenth century had allowed some empirical progress in the study of the weather, but subsequent advances rested firmly on knowledge of the theoretical mechanics of fluids in a rotating frame of reference and on improved understanding of the physics of moist air and of radiation. Admittedly, every big step forward in data acquisition, such as the radiosonde balloon, enlarged the potentialities of meteorology as a science; nevertheless, the data were only useful when interpreted in terms of ideas emanating in part from theoretical mechanics. Some of these ideas proved useful also in the study of planetary atmospheres.

The modern meteorologist is able, however, to go far beyond the simple introduction of ideas from fluid dynamics. The global collection of data from radiosondes and from meteorological satellites now permits initial conditions to be determined that allow a forward numerical integration of the partial differential equations for atmospheric motions (including the necessary radiation physics) to be carried out in finite-difference form and used successfully to calculate weather conditions a few days ahead. This rapidly developing field is one of today's particularly important areas of application of theoretical mechanics.

Similar stories can be told in two other branches of the study of the earth's fluid envelope: Oceanography, with its highly developed study of ocean tides and currents intimately related to hydrodynamic theory; and ionospheric dynamics, including the dynamics of the interaction of the earth's magnetosphere with the solar wind of charged particles emanating from the sun, Increasingly, the mechanics of the entire fluid envelope needs to be treated as a whole.

Solid-earth geophysics is also highly dependent on mechanics and on its well developed seismological theories, allowing information about the interior of the earth to be inferred from data gathered by sensitive seismographs on how waves generated by earthquakes propagate through it. Extensive knowledge of the material properties of the solid outer parts of the earth (the crust and mantle) were derived by these methods; which in addition, demonstrated the existence of the liquid core. Later, it became recognized that the magnetic field of the earth is generated by 'dynamo' motions within the electrically conducting core through mechanisms that in recent times have been in part clarified by magneto-fluid-dynamics. All of these studies of the earth's interior are found valuable also in investigations of the structure of the moon and of the other planets and their satellites.

Life Sciences

This was the period during which, first of all, the very diverse modes of animal locomotion were subjected to detailed analysis from the standpoint of the forces required and of how they are exerted by muscles and other motile organs. That work has comprehensively applied the principles of mechanics of solids to all of the different gaits (including creeping, walking, hopping, trotting, cantering, and galloping) used in terrestrial locomotion, Simultaneously, aquatic locomotion was being studied from the standpoint of hydrodynamics for a huge range of fishes and other vertebrate swimmers, for mobile crustaceans and molluscs, and for flagellar and ciliate microorganisms. Aerodynamics, again, was used to analyse the many and various modes of forward flight and its control, of takeoff and landing, and of climbing, soaring, diving and hovering used by birds, bats, and insects.

The biomechanics of bone and connective tissue was comprehensively investigated, with applications to the design of prosthetic and orthotic devices. Physiological fluid dynamics was developed, too, for the detailed analysis of blood flow in the cardiovascular system, of air flow in the respiratory tract, and of fluid motions in various specialised organs such as the ear. There are, furthermore, many other excellent possibilities for future applications of mechanics in the life sciences.

Engineering Mechanics

Engineering mechanics is understood by many to be the application of mechanics to solve problems. Yet, for engineers, this definition is a bit vague, as it overlooks key questions like; What is being applied? And how exactly are these constitutions working together?

To answer such questions, one must first examine the fundamentals of engineering mechanics, and then see how these fundamentals work together to solve engineering problems.

Forces, torques and dynamics will be some of the defined fundamentals, but the importance of obtaining a holistic understanding both of the elements and their interactions should first be noted. The rationale for obtaining a holistic understanding is that civil engineers will not only be expected to solve problems, but will also need to explain how problems are, or could be, solved.

Most of the time engineers will be explaining to individuals who don't have a firm grasp on the engineering vernacular. Therefore, for clear communication, it will often be necessary to translate engineering knowledge into something more consumable for the general public.

Engineers who possess a fundamental knowledge of engineering mechanics will stand apart from their colleagues, as they will be able to swiftly flow from analyzing with other engineers to explaining potential problems to those not mechanically savvy.

A force is defined as an application of pressure either geared toward or pulling away from an object. The exertion of this force always involves an external – i.e. outside, different – object or source, such as gravity or magnetism.

Mathematically speaking, a force is considered to be an object's mass multiplied by its acceleration ($F = m \cdot a$). It should be noted though, that all objects, even those that don't appear to be moving, have forces acting upon them. Thus, the power and direction of a force determines that of a moving object, whereas it is the counter-force that determines the power and direction of a static object.

Keeping forces in mind, we will next examine torque, which is understood to be the rotational counterpart of a force, and is produced via the application of one or more forces. In particular, torque is often relative to a previously determined pivot or rotational axis.

The fundamentals of torque are predominately dependent upon the force(s) it interacts with. For example, torque is directly influenced by the magnitude of the force, as well as how far the applied force is from the rotation axis. Some real-world examples of torque would include the motion of a golf swing or the rotating blades of a lawnmower.

Statics are understood as a branch of mechanics that analyze the torques and forces acting upon resting bodies. For engineers, these resting bodies are commonly referred to as rigid bodies (a body upon which there is a constant, never-changing distance between two points, no matter the force applied on it).

With rigid bodies comes what engineers describe as static equilibrium. In particular, with statics, engineers are interested in studying the effects and distribution of torques and forces upon this equilibrium.

Dynamics also relate to torques and forces, but this branch of engineering is particularly interested in the impact that torques and forces have on motion. Thus, dynamics could be properly described as the vector of engineering mechanics that study bodies in motion.

The branch of dynamics is segmented into two fundamental parts: kinematics and kinetics. Kinematics involves motion, be it that of a particle or body, yet this particular segment of dynamics is not at all concerned with the forces behind the motion. Conversely, kinetics is concerned with the acting force.

Applied Mechanics

2

- **Vectors**
- **Momentum**
- **Newton's Laws of Motion**
- **Friction**
- **Law of Restitution**
- **Principle of Transmissibility**

Some of the fundamental principles and concepts that fall under the domain of applied mechanics are vectors, momentum, Newton's laws of motion, friction, law of restitution, etc. This chapter has been carefully written to provide an easy understanding of these principle and concepts of applied mechanics.

The science which studies effect of forces on the bodies at rest or in motion is known as Mechanics. There is various knowledge of mechanics required in various field of different subjects like:

- Strength of materials.
- Theory of structure.
- Design of structure.
- Theory and design of mechanics.

The law and principle of mechanics which used or applied in engineering problems in various field, is called as Applied Mechanics.

Mechanics can be further divided in 2 main branches one is 'Static' and other is 'Dynamic'.

- Static: It deals with the system of forces acting on bodies at rest.
- Dynamic: It deals with the effect of forces on bodies in motion.

Dynamic is divided in branches kinetics and kinematics:

- Kinetics: It is the branch of dynamics which studies motion considering mass of body and force causing motion.

- Kinematics: It is the branch of dynamics which studies without considering the force of motion.

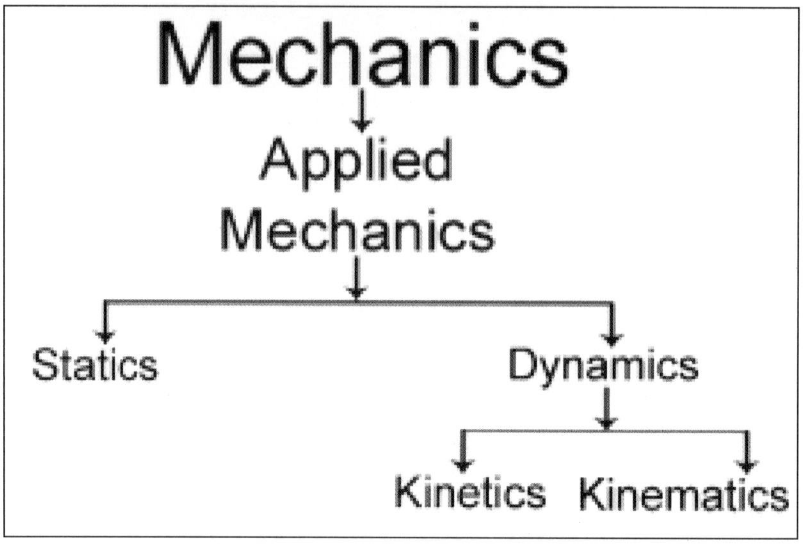

Fundamental concepts of Applied Mechanics

Following are Fundamental Concept of Mechanics:

- Space.
- Mass.
- Time.
- Force.

Space: It is concern with position of a body to fix the position of point.

Mass: To distinguish behaviour of the two bodies under the action of an identical force.

Time: This term mostly used in dynamics to relate the sequence of events.

Force: It is an essential agency which changes or tends to change the state of rest or uniform motion of body.

Fundamental Principles of Applied Mechanics

- Law of parallelogram of force.
- Law of transmissibility.
- Newton's first law of motion.
- Newton's second law of motion.
- Newton's third law of motion.
- Newton's law of universal gravitation.

System of Units

- CGS system: Centimeter-gram-second system.
- FPS system: Foot-pound-second system.
- MKS system: Metre-kilogram-second system.
- SI system: International system of units.

Vectors

The equations of mechanics are typically written in terms of Cartesian coordinates. At a certain time t, the position of a particle may be specified by giving its coordinates x(t), y(t), and z(t) in a particular Cartesian frame of reference. However, a different observer of the same particle might choose a differently oriented set of mutually perpendicular axes, say, x′, y′, and z′. The motion of the particle is then described by the first observer in terms of the rate of change of x(t), y(t), and z(t), while the second observer would discuss the rates of change of x′(t), y′(t), and z′(t). That is, both observers see the same particle executing the same motion and obeying the same laws, but they describe the situation with different equations. This awkward situation may be avoided by means of a mathematical construction called a vector. Although vectors are mathematically simple and extremely useful in discussing mechanics, they were not developed in their modern form until late in the 19th century, when J. Willard Gibbs and Oliver Heaviside (of the United States and Britain, respectively) each applied vector analysis in order to help express the new laws of electromagnetism proposed by James Clerk Maxwell.

A vector is a quantity that has both magnitude and direction. It is typically represented symbolically by an arrow in the proper direction, whose length is proportional to the magnitude of the vector. Although a vector has magnitude and direction, it does not have position. A vector is not altered if it is displaced parallel to itself as long as its length is not changed.

By contrast to a vector, an ordinary quantity having magnitude but not direction is known as a scalar. In printed works vectors are often represented by boldface letters such as A or X, and scalars are represented by lightface letters, A or X. The magnitude of a vector, denoted $|A|$, is itself a scalar—i.e., $|A| = A$.

Because vectors are different from ordinary (i.e., scalar) quantities, all mathematical operations involving vectors must be carefully defined. There is no mathematical operation that corresponds to division by a vector.

If vector A is added to vector B, the result is another vector, C, written A + B = C. The operation is performed by displacing B so that it begins where A ends, as shown in figure. C is then the vector that starts where A begins and ends where B ends.

Vector addition is defined to have the (nontrivial) property A + B = B + A. There do exist quantities having magnitude and direction that do not obey this requirement. An example is finite rotations in space. Two finite rotations of a body about different axes do not necessarily result in the same orientation if performed in the opposite order.

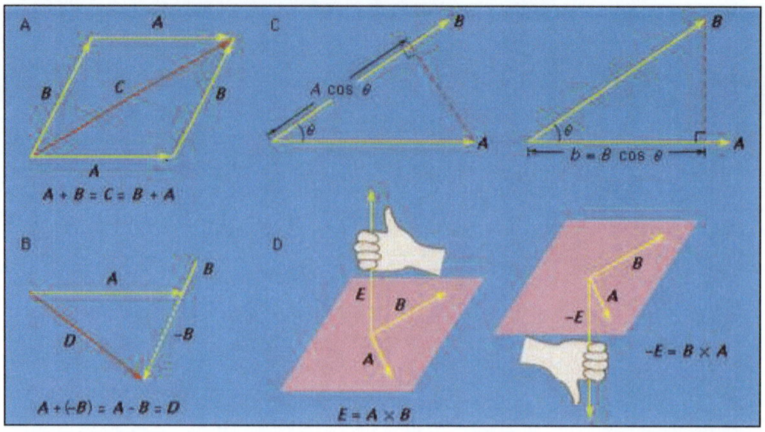

(A) The vector sum C = A + B = B + A. (B) The vector difference A + (–B) = A – B = D. (C, left) A cos θ is the component of A along B and (right) B cos θ is the component of B along A. (D, left) The right-hand rule used to find the direction of E = A × B and (right) the right-hand rule used to find the direction of –E = B × A.

Vector subtraction is defined by A – B = A + (–B), where the vector –B has the same magnitude as B but the opposition direction. The idea is illustrated in figure.

A vector may be multiplied by a scalar. Thus, for example, the vector 2A has the same direction as A but is twice as long. If the scalar has dimensions, the resulting vector still has the same direction as the original one, but the two cannot be compared in magnitude. For example, a particle moving with constant velocity v suffers a displacement s in time t given by s = vt. The vector v has been multiplied by the scalar t to give a new vector, s, which has the same direction as v but cannot be compared to v in magnitude (a displacement of one metre is neither bigger nor smaller than a velocity of one metre per second). This is a typical example of a phenomenon that might be represented by different equations in differently oriented Cartesian coordinate systems but that has a single vector equation (for all observers not moving with respect to one another).

The dot product (also known as the scalar product, or sometimes the inner product) is an operation that combines two vectors to form a scalar. The operation is written A · B. If θ is the (smaller) angle between A and B, then the result of the operation is A · B = AB cos θ. The dot product measures the extent to which two vectors are parallel. It may be thought of as multiplying the magnitude of one vector (either one) by the projection of the other upon it, as shown in figure. If the two vectors are perpendicular, the dot product is zero.

The cross product (also known as the vector product) combines two vectors to form another vector, perpendicular to the plane of the original vectors. The operation is written A × B. If θ is the (smaller) angle between A and B, then|A × B|= AB sin θ. The direction of A × B is given by the right-hand rule: if the fingers of the right hand are made to rotate from A through θ to B, the thumb points in the direction of A × B, as shown in figure. The cross product is zero if the two vectors are parallel, and it is maximum in magnitude if they are perpendicular.

The derivative, or rate of change, of a vector is defined in perfect analogy to the derivative of a scalar: if the vector A changes with time t, then:

$$\frac{dA}{dt} = \lim_{x \to \infty} \frac{A(t+\Delta t) - A(t)}{\Delta t}.$$

Before going to the limit on the right-hand side of equation, the operations described are vector subtraction [A(t + Δt) – A(t)] and scalar multiplication (by 1/Δt). The result, dA/dt, is therefore itself a vector. Notice that, as shown in figure, the difference between two vectors, in this case A(t + Δt) – A(t), may be in quite a different direction than either of the vectors from which it is formed, here A(t + Δt) and A(t). As a result, dA/dt may be in a different direction than A(t).

Momentum

Linear momentum, translational momentum, or simply momentum (pl. momenta) is the product of the mass and velocity of an object. It is a vector quantity, possessing a magnitude and a direction. If m is an object's mass and v is its velocity (also a vector quantity), then the object's momentum is:

$$p = mv.$$

In SI units, momentum is measured in kilogram meters per second (kg, m/s).

Newton's second law of motion states that a body's rate of change in momentum is equal to the net force acting on it. Momentum depends on the frame of reference, but in any inertial frame it is a *conserved* quantity, meaning that if a closed system is not affected by external forces, its total linear momentum does not change. Momentum is also conserved in special relativity (with a modified formula) and, in a modified form, in electrodynamics, quantum mechanics, quantum field theory, and general relativity. It is an expression of one of the fundamental symmetries of space and time: translational symmetry.

Advanced formulations of classical mechanics, Lagrangian and Hamiltonian mechanics, allow one to choose coordinate systems that incorporate symmetries and constraints. In these systems the conserved quantity is *generalized momentum*, and in general this is different from the *kinetic* momentum defined above. The concept of generalized momentum is carried over into quantum mechanics, where it becomes an operator on a wave function. The momentum and position operators are related by the Heisenberg uncertainty principle.

In continuous systems such as electromagnetic fields, fluids and deformable bodies, a momentum density can be defined, and a continuum version of the conservation of momentum leads to equations such as the Navier–Stokes equations for fluids or the Cauchy momentum equation for deformable solids or fluids.

Newtonian

Momentum is a vector quantity: It has both magnitude and direction. Since momentum has a direction, it can be used to predict the resulting direction and speed of motion of objects after they collide. The vector equations are almost identical to the scalar equations.

Single Particle

The momentum of a particle is conventionally represented by the letter p. It is the product of two quantities, the particle's mass (represented by the letter m) and its velocity (v):

$$p = mv.$$

The unit of momentum is the product of the units of mass and velocity. In SI units, if the mass is in kilograms and the velocity is in meters per second then the momentum is in kilogram meters per second (kg·m/s). In cgs units, if the mass is in grams and the velocity in centimeters per second, then the momentum is in gram centimeters per second (g·cm/s).

Being a vector, momentum has magnitude and direction. For example, a 1 kg model airplane, traveling due north at 1 m/s in straight and level flight, has a momentum of 1 kg·m/s due north measured with reference to the ground.

Many Particles

The momentum of a system of particles is the vector sum of their momenta. If two particles have respective masses m_1 and m_2, and velocities v_1 and v_2, the total momentum is:

$$\begin{aligned} p &= p_1 + p_2 \\ &= m_1 v_1 + m_2 v_2 . \end{aligned}$$

The momenta of more than two particles can be added more generally with the following:

$$p = \sum_i m_i v_i .$$

A system of particles has a center of mass, a point determined by the weighted sum of their positions:

$$r_{\mathrm{cm}} = \frac{m_1 r_1 + m_2 r_2 + \cdots}{m_1 + m_2 + \cdots} = \frac{\sum_i m_i r_i}{\sum_i m_i} .$$

If one or more of the particles is moving, the center of mass of the system will generally be moving as well (unless the system is in pure rotation around it). If the total mass of the particles is m, and the center of mass is moving at velocity v_{cm}, the momentum of the system is:

$$p = m v_{\mathrm{cm}} .$$

This is known as Euler's first law.

Relation to Force

If the net force F applied to a particle is constant, and is applied for a time interval Δt, the momentum of the particle changes by an amount:

$$\Delta p = F \Delta t .$$

In differential form, this is Newton's second law; the rate of change of the momentum of a particle is equal to the instantaneous force F acting on it,

$$F = \frac{dp}{dt} .$$

If the net force experienced by a particle changes as a function of time, *F(t)*, the change in momentum (or impulse J) between times t_1 and t_2 is:

$$\Delta p = J = \int_{t_1}^{t_2} F(t)\, dt.$$

Impulse is measured in the derived units of the newton second (1 N·s = 1 kg·m/s) or dyne second (1 dyne·s = 1 g·cm/s)

Under the assumption of constant mass m, it is equivalent to write

$$F = \frac{d(mv)}{dt} = m\frac{dv}{dt} = ma,$$

Hence the net force is equal to the mass of the particle times its acceleration.

Example: A model airplane of mass 1 kg accelerates from rest to a velocity of 6 m/s due north in 2 s. The net force required to produce this acceleration is 3 newtons due north. The change in momentum is 6 kg·m/s due north. The rate of change of momentum is 3 (kg·m/s)/s due north which is numerically equivalent to 3 newtons.

Conservation

In a closed system (one that does not exchange any matter with its surroundings and is not acted on by external forces) the total momentum is constant. This fact, known as the *law of conservation of momentum*, is implied by Newton's laws of motion. Suppose, for example, that two particles interact. Because of the third law, the forces between them are equal and opposite. If the particles are numbered 1 and 2, the second law states that $F_1 = \frac{dp_1}{dt}$ *and* $F_2 = \frac{dp_2}{dt}$. Therefore,

$$\frac{dp_1}{dt} = -\frac{dp_2}{dt},$$

with the negative sign indicating that the forces oppose. Equivalently,

$$\frac{d}{dt}(p_1 + p_2) = 0.$$

If the velocities of the particles are u_1 and u_2 before the interaction, and afterwards they are v_1 and v_2, then:

$$m_1u_1 + m_2u_2 = m_1v_1 + m_2v_2.$$

This law holds no matter how complicated the force is between particles. Similarly, if there are several particles, the momentum exchanged between each pair of particles adds up to zero, so the total change in momentum is zero. This conservation law applies to all interactions, including collisions and separations caused by explosive forces. It can also be generalized to situations where Newton's laws do not hold, for example in the theory of relativity and in electrodynamics.

Dependence on Reference Frame

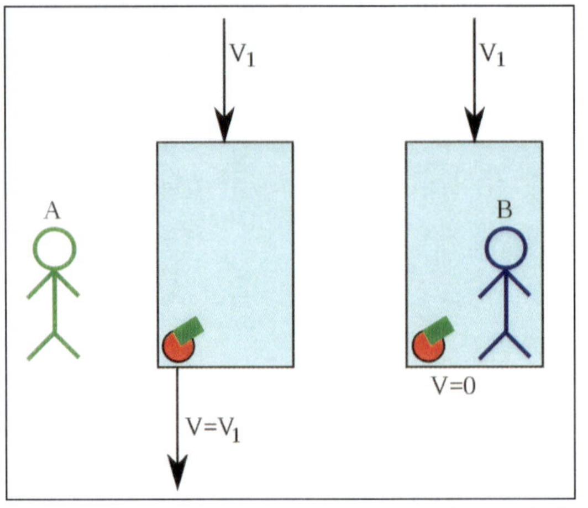

Newton's apple in Einstein's elevator. In person A's frame of reference, the apple has non-zero velocity and momentum. In the elevator's and person B's frames of reference, it has zero velocity and momentum.

Momentum is a measurable quantity, and the measurement depends on the motion of the observer. For example: if an apple is sitting in a glass elevator that is descending, an outside observer, looking into the elevator, sees the apple moving, so, to that observer, the apple has a non-zero momentum. To someone inside the elevator, the apple does not move, so, it has zero momentum. The two observers each have a frame of reference, in which, they observe motions, and, if the elevator is descending steadily, they will see behavior that is consistent with those same physical laws.

Suppose a particle has position x in a stationary frame of reference. From the point of view of another frame of reference, moving at a uniform speed u, the position (represented by a primed coordinate) changes with time as:

$$x' = x - ut\,.$$

This is called a Galilean transformation. If the particle is moving at speed $\frac{dx}{dt} = v$ in the first frame of reference, in the second, it is moving at speed:

$$v' = \frac{dx'}{dt} = v - u\,.$$

Since u does not change, the accelerations are the same:

$$a' = \frac{dv'}{dt} = a\,.$$

Thus, momentum is conserved in both reference frames. Moreover, as long as the force has the same form, in both frames, Newton's second law is unchanged. Forces such as Newtonian gravity, which depend only on the scalar distance between objects, satisfy this criterion. This independence of reference frame is called Newtonian relativity or Galilean invariance.

A change of reference frame, can, often, simplify calculations of motion. For example, in a collision of two particles, a reference frame can be chosen, where, one particle begins at rest. Another, commonly used reference frame, is the center of mass frame – one that is moving with the center of mass. In this frame, the total momentum is zero.

Application to Collisions

By itself, the law of conservation of momentum is not enough to determine the motion of particles after a collision. Another property of the motion, kinetic energy, must be known. This is not necessarily conserved. If it is conserved, the collision is called an *elastic collision*; if not, it is an *inelastic collision*.

Elastic Collisions

An elastic collision is one in which no kinetic energy is absorbed in the collision. Perfectly elastic "collisions" can occur when the objects do not touch each other, as for example in atomic or nuclear scattering where electric repulsion keeps them apart. A slingshot maneuver of a satellite around a planet can also be viewed as a perfectly elastic collision. A collision between two pool balls is a good example of an *almost* totally elastic collision, due to their high rigidity, but when bodies come in contact there is always some dissipation.

A head-on elastic collision between two bodies can be represented by velocities in one dimension, along a line passing through the bodies. If the velocities are u_1 and u_2 before the collision and v_1 and v_2 after, the equations expressing conservation of momentum and kinetic energy are:

$$m_1u_1 + m_2u_2 = m_1v_1 + m_2v_2$$
$$\tfrac{1}{2}m_1u_1^2 + \tfrac{1}{2}m_2u_2^2 = \tfrac{1}{2}m_1v_1^2 + \tfrac{1}{2}m_2v_2^2 .$$

A change of reference frame can simplify analysis of a collision. For example, suppose there are two bodies of equal mass m, one stationary and one approaching the other at a speed v. The center of mass is moving at speed $\frac{v}{2}$ and both bodies are moving towards it at speed $\frac{v}{2}$. Because of the symmetry, after the collision both must be moving away from the center of mass at the same speed. Adding the speed of the center of mass to both, we find that the body that was moving is now stopped and the other is moving away at speed v. The bodies have exchanged their velocities. Regardless of the velocities of the bodies, a switch to the center of mass frame leads us to the same conclusion. Therefore, the final velocities are given by:

$$v_1 = u_2$$
$$v_2 = u_1 .$$

In general, when the initial velocities are known, the final velocities are given by:

$$v_1 = \left(\frac{m_1 - m_2}{m_1 + m_2}\right)u_1 + \left(\frac{2m_2}{m_1 + m_2}\right)u_2$$

$$v_2 = \left(\frac{m_2 - m_1}{m_1 + m_2}\right)u_2 + \left(\frac{2m_1}{m_1 + m_2}\right)u_1 .$$

If one body has much greater mass than the other, its velocity will be little affected by a collision while the other body will experience a large change.

Inelastic Collisions

In an inelastic collision, some of the kinetic energy of the colliding bodies is converted into other forms of energy (such as heat or sound). Examples include traffic collisions, in which the effect of loss of kinetic energy can be seen in the damage to the vehicles; electrons losing some of their energy to atoms (as in the Franck–Hertz experiment); and particle accelerators in which the kinetic energy is converted into mass in the form of new particles.

In a perfectly inelastic collision (such as a bug hitting a windshield), both bodies have the same motion afterwards. If one body is motionless to begin with, the equation for conservation of momentum is:

$$m_1 u_1 = (m_1 + m_2) v,$$

So

$$v = \frac{m_1}{m_1 + m_2} u_1 .$$

In a frame of reference moving at the speed v), the objects are brought to rest by the collision and 100% of the kinetic energy is converted to other forms of energy.

One measure of the inelasticity of the collision is the coefficient of restitution C_R, defined as the ratio of relative velocity of separation to relative velocity of approach. In applying this measure to a ball bouncing from a solid surface, this can be easily measured using the following formula:

$$C_R = \sqrt{\frac{\text{bounce height}}{\text{drop height}}}.$$

The momentum and energy equations also apply to the motions of objects that begin together and then move apart. For example, an explosion is the result of a chain reaction that transforms potential energy stored in chemical, mechanical, or nuclear form into kinetic energy, acoustic energy, and electromagnetic radiation. Rockets also make use of conservation of momentum: propellant is thrust outward, gaining momentum, and an equal and opposite momentum is imparted to the rocket.

Multiple Dimensions

Real motion has both direction and velocity and must be represented by a vector. In a coordinate system with x, y, z axes, velocity has components v_x in the x-direction, v_y in the y-direction, v_z in the z-direction. The vector is represented by a boldface symbol:

$$\mathbf{v} = (v_x, v_y, v_z).$$

Similarly, the momentum is a vector quantity and is represented by a boldface symbol:

$$\mathbf{p} = \left(p_x, p_y, p_z\right).$$

Work in vector form if the scalars p and v are replaced by vectors p and v. Each vector equation represents three scalar equations. For example,

$$\mathbf{p} = m\mathbf{v}$$

represents three equations:

$$p_x = mv_x$$
$$p_y = mv_y$$
$$p_z = mv_z.$$

The kinetic energy equations are exceptions to the above replacement rule. The equations are still one-dimensional, but each scalar represents the magnitude of the vector, for example,

$$v^2 = v_x^2 + v_y^2 + v_z^2.$$

Each vector equation represents three scalar equations. Often coordinates can be chosen so that only two components are needed. Each component can be obtained separately and the results combined to produce a vector result.

A simple construction involving the center of mass frame can be used to show that if a stationary elastic sphere is struck by a moving sphere, the two will head off at right angles after the collision.

Objects of Variable Mass

The concept of momentum plays a fundamental role in explaining the behavior of variable-mass objects such as a rocket ejecting fuel or a star accreting gas. In analyzing such an object, one treats the object's mass as a function that varies with time: m(t). The momentum of the object at time t is therefore p(t) = m(t)v(t). One might then try to invoke Newton's second law of motion by saying that the external force F on the object is related to its momentum p(t) by $F = \frac{dp}{dt}$, but this is incorrect, as is the related expression found by applying the product rule to $\frac{d(mv)}{dt}$:

$$F = m(t)\frac{dv}{dt} + v(t)\frac{dm}{dt}.\text{(incorrect)}$$

This equation does not correctly describe the motion of variable-mass objects. The correct equation is:

$$F = m(t)\frac{dv}{dt} - u\frac{dm}{dt},$$

where u is the velocity of the ejected/accreted mass as seen in the object's rest frame. This is distinct from v, which is the velocity of the object itself as seen in an inertial frame.

This equation is derived by keeping track of both the momentum of the object as well as the momentum of the ejected/accreted mass (*dm*). When considered together, the object and the mass (*dm*) constitute a closed system in which total momentum is conserved.

$$P(t+dt)=(m-dm)(v+dv)+dm(v-u)=mv+mdv-udm=P(t)+mdv-udm$$

Relativistic

Lorentz Invariance

Newtonian physics assumes that absolute time and space exist outside of any observer; this gives rise to Galilean invariance. It also results in a prediction that the speed of light can vary from one reference frame to another. This is contrary to observation. In the special theory of relativity, Einstein keeps the postulate that the equations of motion do not depend on the reference frame, but assumes that the speed of light *c* is invariant. As a result, position and time in two reference frames are related by the Lorentz transformation instead of the Galilean transformation.

Consider, for example, one reference frame moving relative to another at velocity v in the x direction. The Galilean transformation gives the coordinates of the moving frame as:

$$t' = t$$
$$x' = x - vt$$

while the Lorentz transformation gives:

$$t' = \gamma\left(t - \frac{vx}{c^2}\right)$$
$$x' = \gamma\left(x - vt\right)$$

where γ is the Lorentz factor:

$$\gamma = \frac{1}{\sqrt{1 - v^2 / c^2}}.$$

Newton's second law, with mass fixed, is not invariant under a Lorentz transformation. However, it can be made invariant by making the inertial mass m of an object a function of velocity:

$$m = \gamma m_0;$$

m_0 is the object's invariant mass.

The modified momentum,

$$\mathbf{p} = \gamma m_0 \mathbf{v},$$

obeys Newton's second law:

$$\mathbf{F} = \frac{d\mathbf{p}}{dt}.$$

Within the domain of classical mechanics, relativistic momentum closely approximates Newtonian momentum: at low velocity, $\gamma m_0 \mathrm{v}$ is approximately equal to $m_0 \mathrm{v}$, the Newtonian expression for momentum.

Four-vector Formulation

In the theory of special relativity, physical quantities are expressed in terms of four-vectors that include time as a fourth coordinate along with the three space coordinates. These vectors are generally represented by capital letters, for example R for position. The expression for the *four-momentum* depends on how the coordinates are expressed. Time may be given in its normal units or multiplied by the speed of light so that all the components of the four-vector have dimensions of length. If the latter scaling is used, an interval of proper time, τ, defined by:

$$c^2 d\tau^2 = c^2 dt^2 - dx^2 - dy^2 - dz^2 ,$$

is invariant under Lorentz transformations. Mathematically this invariance can be ensured in one of two ways: by treating the four-vectors as Euclidean vectors and multiplying time by $\sqrt{-1}$; or by keeping time a real quantity and embedding the vectors in a Minkowski space. In a Minkowski space, the scalar product of two four-vectors $\mathbf{U} = (U_0, U_1, U_2, U_3)$ and $\mathbf{V} = (V_0, V_1, V_2, V_3)$ is defined as:

$$\mathbf{U} \cdot \mathbf{V} = U_0 V_0 - U_1 V_1 - U_2 V_2 - U_3 V_3 .$$

In all the coordinate systems, the (contravariant) relativistic four-velocity is defined by:

$$\mathbf{U} \equiv \frac{d\mathbf{R}}{d\tau} = \gamma \frac{d\mathbf{R}}{dt},$$

and the (contravariant) four-momentum is:

$$\mathbf{P} = m_0 \mathbf{U},$$

where m_0 is the invariant mass. If R = (ct,x,y,z) (in Minkowski space), then,

$$\mathbf{P} = \gamma m_0 (c, \mathbf{v}) = (mc, \mathbf{p}).$$

Using Einstein's mass-energy equivalence, $E = mc^2$, this can be rewritten as:

$$\mathbf{P} = \left(\frac{E}{c}, \mathbf{p} \right).$$

Thus, conservation of four-momentum is Lorentz-invariant and implies conservation of both mass and energy.

The magnitude of the momentum four-vector is equal to $m_0 c$:

$$\| \mathbf{P} \|^2 = \mathbf{P} \cdot \mathbf{P} = \gamma^2 m_0^2 (c^2 - v^2) = (m_0 c)^2 ,$$

and is invariant across all reference frames.

The relativistic energy–momentum relationship holds even for massless particles such as photons; by setting $m_0 = 0$ it follows that:

$$E = pc.$$

In a game of relativistic "billiards", if a stationary particle is hit by a moving particle in an elastic collision, the paths formed by the two afterwards will form an acute angle. This is unlike the non-relativistic case where they travel at right angles.

The four-momentum of a planar wave can be related to a wave four-vector.

$$\mathbf{P} = \left(\frac{E}{c}, \vec{\mathbf{p}}\right) = \hbar\mathbf{K} = \hbar\left(\frac{\omega}{c}, \vec{\mathbf{k}}\right)$$

For a particle, the relationship between temporal components, $E = \hbar\,\omega$, is the Planck–Einstein relation, and the relation between spatial components, $p = \hbar\,k$, describes a de Broglie matter wave.

Generalized

Newton's laws can be difficult to apply to many kinds of motion because the motion is limited by *constraints*. For example, a bead on an abacus is constrained to move along its wire and a pendulum bob is constrained to swing at a fixed distance from the pivot. Many such constraints can be incorporated by changing the normal Cartesian coordinates to a set of *generalized coordinates* that may be fewer in number. Refined mathematical methods have been developed for solving mechanics problems in generalized coordinates. They introduce a *generalized momentum*, also known as the *canonical* or *conjugate momentum*, that extends the concepts of both linear momentum and angular momentum. To distinguish it from generalized momentum, the product of mass and velocity is also referred to as *mechanical*, *kinetic* or *kinematic momentum*.

Lagrangian Mechanics

In Lagrangian mechanics, a Lagrangian is defined as the difference between the kinetic energy T and the potential energy V:

$$\mathcal{L} = T - V.$$

If the generalized coordinates are represented as a vector $\mathbf{q} = (q_1, q_2, \dots, q_N)$ and time differentiation is represented by a dot over the variable, then the equations of motion (known as the Lagrange or Euler–Lagrange equations) are a set of N equations:

$$\frac{d}{dt}\left(\frac{\partial \mathcal{L}}{\partial \dot{q}_j}\right) - \frac{\partial \mathcal{L}}{\partial q_j} = 0.$$

If a coordinate q_i is not a Cartesian coordinate, the associated generalized momentum component p_i does not necessarily have the dimensions of linear momentum. Even if q_i is a Cartesian coordinate, p_i will not be the same as the mechanical momentum if the potential depends on velocity. Some sources represent the kinematic momentum by the symbol Π.

In this mathematical framework, a generalized momentum is associated with the generalized coordinates. Its components are defined as:

$$p_j = \frac{\partial \mathcal{L}}{\partial \dot{q}_j}.$$

Each component p_j is said to be the *conjugate momentum* for the coordinate q_j.

Now if a given coordinate q_i does not appear in the Lagrangian (although its time derivative might appear), then:

$$p_j = \text{constant}.$$

This is the generalization of the conservation of momentum.

Even if the generalized coordinates are just the ordinary spatial coordinates, the conjugate momenta are not necessarily the ordinary momentum coordinates.

Hamiltonian Mechanics

In Hamiltonian mechanics, the Lagrangian (a function of generalized coordinates and their derivatives) is replaced by a Hamiltonian that is a function of generalized coordinates and momentum. The Hamiltonian is defined as:

$$\mathcal{H}(\mathbf{q},\mathbf{p},t) = \mathbf{p}\cdot\dot{\mathbf{q}} - \mathcal{L}(\mathbf{q},\dot{\mathbf{q}},t),$$

where the momentum is obtained by differentiating the Lagrangian as above. The Hamiltonian equations of motion are:

$$\dot{q}_i = \frac{\partial \mathcal{H}}{\partial p_i}$$

$$-\dot{p}_i = \frac{\partial \mathcal{H}}{\partial q_i}$$

$$-\frac{\partial \mathcal{L}}{\partial t} = \frac{d\mathcal{H}}{dt}.$$

As in Lagrangian mechanics, if a generalized coordinate does not appear in the Hamiltonian, its conjugate momentum component is conserved.

Symmetry and Conservation

Conservation of momentum is a mathematical consequence of the homogeneity (shift symmetry) of space (position in space is the canonical conjugate quantity to momentum). That is, conservation of momentum is a consequence of the fact that the laws of physics do not depend on position; this is a special case of Noether's theorem.

Electromagnetic

Particle in a Field

In Maxwell's equations, the forces between particles are mediated by electric and magnetic fields. The electromagnetic force (*Lorentz force*) on a particle with charge q due to a combination of electric field E and magnetic field B is:

$$\mathbf{F} = q(\mathbf{E} + \mathbf{v} \times \mathbf{B})$$

(in SI units). It has an electric potential $\varphi(\mathrm{r}, t)$ and magnetic vector potential $\mathrm{A}(\mathrm{r}, t)$. In the non-relativistic regime, its generalized momentum is:

$$\mathbf{P} = m\mathbf{v} + q\mathbf{A},$$

while in relativistic mechanics this becomes:

$$\mathbf{P} = \gamma m\mathbf{v} + q\mathbf{A}.$$

Conservation

In Newtonian mechanics, the law of conservation of momentum can be derived from the law of action and reaction, which states that every force has a reciprocating equal and opposite force. Under some circumstances, moving charged particles can exert forces on each other in non-opposite directions. Nevertheless, the combined momentum of the particles and the electromagnetic field is conserved.

Vacuum

The Lorentz force imparts a momentum to the particle, so by Newton's second law the particle must impart a momentum to the electromagnetic fields.

In a vacuum, the momentum per unit volume is:

$$\mathbf{g} = \frac{1}{\mu_0 c^2}\mathbf{E} \times \mathbf{B},$$

where μ_o is the vacuum permeability and c is the speed of light. The momentum density is proportional to the Poynting vector S which gives the directional rate of energy transfer per unit area:

$$\mathbf{g} = \frac{\mathbf{S}}{c^2}.$$

If momentum is to be conserved over the volume V over a region Q, changes in the momentum of matter through the Lorentz force must be balanced by changes in the momentum of the electromagnetic field and outflow of momentum. If $\mathrm{P}_{\mathrm{mech}}$ is the momentum of all the particles in Q, and the particles are treated as a continuum, then Newton's second law gives:

$$\frac{d\mathbf{P}_{\mathrm{mech}}}{dt} = \iiint\limits_{Q} (\rho\mathbf{E} + \mathbf{J} \times \mathbf{B})\, dV.$$

The electromagnetic momentum is:

$$\mathbf{P}_{\text{field}} = \frac{1}{\mu_0 c^2} \iiint_Q \mathbf{E} \times \mathbf{B}\, dV,$$

and the equation for conservation of each component i of the momentum is:

$$\frac{d}{dt}\left(\mathbf{P}_{\text{mech}} + \mathbf{P}_{\text{field}}\right)_i = \iint_\sigma \left(\sum_j T_{ij} n_j\right) d\Sigma.$$

The term on the right is an integral over the surface area Σ of the surface σ representing momentum flow into and out of the volume, and n_j is a component of the surface normal of S. The quantity T_{ij} is called the Maxwell stress tensor, defined as:

$$T_{ij} \equiv \epsilon_0 \left(E_i E_j - \frac{1}{2}\delta_{ij} E^2 \right) + \frac{1}{\mu_0}\left(B_i B_j - \frac{1}{2}\delta_{ij} B^2 \right).$$

Media

The above results are for the microscopic Maxwell equations, applicable to electromagnetic forces in a vacuum (or on a very small scale in media). It is more difficult to define momentum density in media because the division into electromagnetic and mechanical is arbitrary. The definition of electromagnetic momentum density is modified to:

$$\mathbf{g} = \frac{1}{c^2}\mathbf{E} \times \mathbf{H} = \frac{\mathbf{S}}{c^2},$$

where the H-field H is related to the B-field and the magnetization M by:

$$\mathbf{B} = \mu_0 \left(\mathbf{H} + \mathbf{M}\right).$$

The electromagnetic stress tensor depends on the properties of the media.

Quantum Mechanical

In quantum mechanics, momentum is defined as a self-adjoint operator on the wave function. The Heisenberg uncertainty principle defines limits on how accurately the momentum and position of a single observable system can be known at once. In quantum mechanics, position and momentum are conjugate variables.

For a single particle described in the position basis the momentum operator can be written as:

$$\mathbf{p} = \frac{\hbar}{i}\nabla = -i\hbar\nabla,$$

where ∇ is the gradient operator, $\hbar$ is the reduced Planck constant, and i is the imaginary unit. This is a commonly encountered form of the momentum operator, though the momentum operator in other bases can take other forms. For example, in momentum space the momentum operator is represented as:

$$\mathbf{p}\psi(p) = p\psi(p),$$

where the operator p acting on a wave function $\psi(p)$ yields that wave function multiplied by the value p, in an analogous fashion to the way that the position operator acting on a wave function $\psi(x)$ yields that wave function multiplied by the value x.

For both massive and massless objects, relativistic momentum is related to the phase constant β by:

$$p = \hbar\beta$$

Electromagnetic radiation (including visible light, ultraviolet light, and radio waves) is carried by photons. Even though photons (the particle aspect of light) have no mass, they still carry momentum. This leads to applications such as the solar sail. The calculation of the momentum of light within dielectric media is somewhat controversial.

In Deformable Bodies and Fluids

Conservation in a Continuum

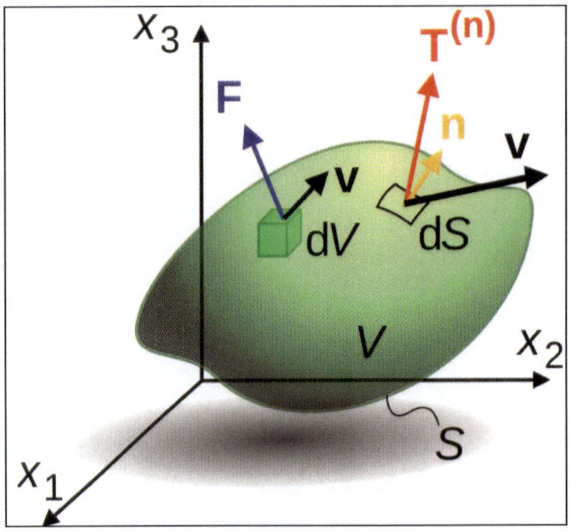

Motion of a material body.

In fields such as fluid dynamics and solid mechanics, it is not feasible to follow the motion of individual atoms or molecules. Instead, the materials must be approximated by a continuum in which there is a particle or fluid parcel at each point that is assigned the average of the properties of atoms in a small region nearby. In particular, it has a density ρ and velocity v that depend on time t and position r. The momentum per unit volume is ρv.

Consider a column of water in hydrostatic equilibrium. All the forces on the water are in balance and the water is motionless. On any given drop of water, two forces are balanced. The first is gravity, which acts directly on each atom and molecule inside. The gravitational force per unit volume is ρg, where g is the gravitational acceleration. The second force is the sum of all the forces exerted on its surface by the surrounding water. The force from below is greater than the force from above by just the amount needed to balance gravity. The normal force per unit area is the pressure p. The average force per unit volume inside the droplet is the gradient of the pressure, so the force balance equation is:

$$-\nabla p + \rho \mathbf{g} = 0.$$

If the forces are not balanced, the droplet accelerates. This acceleration is not simply the partial derivative $\frac{\partial v}{\partial t}$ because the fluid in a given volume changes with time. Instead, the material derivative is needed:

$$\frac{D}{Dt} \equiv \frac{\partial}{\partial t} + \mathbf{v} \cdot \nabla .$$

Applied to any physical quantity, the material derivative includes the rate of change at a point and the changes due to advection as fluid is carried past the point. Per unit volume, the rate of change in momentum is equal to $\rho \frac{Dv}{Dt}$. This is equal to the net force on the droplet.

Forces that can change the momentum of a droplet include the gradient of the pressure and gravity, as above. In addition, surface forces can deform the droplet. In the simplest case, a shear stress τ, exerted by a force parallel to the surface of the droplet, is proportional to the rate of deformation or strain rate. Such a shear stress occurs if the fluid has a velocity gradient because the fluid is moving faster on one side than another. If the speed in the x direction varies with z, the tangential force in direction x per unit area normal to the z direction is:

$$\sigma_{zx} = -\mu \frac{\partial v_x}{\partial z},$$

where μ is the viscosity. This is also a flux, or flow per unit area, of x-momentum through the surface.

Including the effect of viscosity, the momentum balance equations for the incompressible flow of a Newtonian fluid are:

$$\rho \frac{D\mathbf{v}}{Dt} = -\nabla p + \mu \nabla^2 \mathbf{v} + \rho \mathbf{g}.$$

These are known as the Navier–Stokes equations.

The momentum balance equations can be extended to more general materials, including solids. For each surface with normal in direction i and force in direction j, there is a stress component σ_{ij}. The nine components make up the Cauchy stress tensor σ, which includes both pressure and shear. The local conservation of momentum is expressed by the Cauchy momentum equation:

$$\rho \frac{D\mathbf{v}}{Dt} = \nabla \cdot \sigma + \mathbf{f},$$

where f is the body force.

The Cauchy momentum equation is broadly applicable to deformations of solids and liquids. The relationship between the stresses and the strain rate depends on the properties of the material.

Acoustic Waves

A disturbance in a medium gives rise to oscillations, or waves, that propagate away from their source. In a fluid, small changes in pressure p can often be described by the acoustic wave equation:

$$\frac{\partial^2 p}{\partial t^2} = c^2 \nabla^2 p,$$

Where c is the speed of sound. In a solid, similar equations can be obtained for propagation of pressure (P-waves) and shear (S-waves).

The flux, or transport per unit area, of a momentum component ρv_j by a velocity v_i is equal to ρ $vjvj$. In the linear approximation that leads to the above acoustic equation, the time average of this flux is zero. However, nonlinear effects can give rise to a nonzero average. It is possible for momentum flux to occur even though the wave itself does not have a mean momentum.

Newton's Laws of Motion

Newton's laws of motion are three physical laws that, together, laid the foundation for classical mechanics. They describe the relationship between a body and the forces acting upon it, and its motion in response to those forces. More precisely, the first law defines the force qualitatively, the second law offers a quantitative measure of the force, and the third asserts that a single isolated force doesn't exist. These three laws have been expressed in several ways, over nearly three centuries, and can be summarised as follows:

First law:	In an inertial frame of reference, an object either remains at rest or continues to move at a constant velocity, unless acted upon by a force.
Second law:	In an inertial frame of reference, the vector sum of the forces F on an object is equal to the mass m of that object multiplied by the acceleration a of the object: F = ma. (It is assumed here that the mass m is constant)
Third law:	When one body exerts a force on a second body, the second body simultaneously exerts a force equal in magnitude and opposite in direction on the first body.

The three laws of motion were first compiled by Isaac Newton in his Philosophiæ Naturalis Principia Mathematica (Mathematical Principles of Natural Philosophy), first published in 1687. Newton used them to explain and investigate the motion of many physical objects and systems. For example, in the third volume of the text, Newton showed that these laws of motion, combined with his law of universal gravitation, explained Kepler's laws of planetary motion.

Some also describe a fourth law which states that forces add up like vectors, that is, that forces obey the principle of superposition.

Isaac Newton, the physicist who formulated the laws.

Newton's laws are applied to objects which are idealised as single point masses, in the sense that the size and shape of the object's body are neglected to focus on its motion more easily. This can be done when the object is small compared to the distances involved in its analysis, or the deformation and rotation of the body are of no importance. In this way, even a planet can be idealised as a particle for analysis of its orbital motion around a star.

In their original form, Newton's laws of motion are not adequate to characterise the motion of rigid bodies and deformable bodies. Leonhard Euler in 1750 introduced a generalisation of Newton's laws of motion for rigid bodies called Euler's laws of motion, later applied as well for deformable bodies assumed as a continuum. If a body is represented as an assemblage of discrete particles, each governed by Newton's laws of motion, then Euler's laws can be derived from Newton's laws. Euler's laws can, however, be taken as axioms describing the laws of motion for extended bodies, independently of any particle structure.

Newton's laws hold only with respect to a certain set of frames of reference called Newtonian or inertial reference frames. Some authors interpret the first law as defining what an inertial reference frame is; from this point of view, the second law holds only when the observation is made from an inertial reference frame, and therefore the first law cannot be proved as a special case of the second. Other authors do treat the first law as a corollary of the second. The explicit concept of an inertial frame of reference was not developed until long after Newton's death.

In the given interpretation mass, acceleration, momentum, and (most importantly) force are assumed to be externally defined quantities. This is the most common, but not the only interpretation of the way one can consider the laws to be a definition of these quantities.

Newtonian mechanics has been superseded by special relativity, but it is still useful as an approximation when the speeds involved are much slower than the speed of light.

Laws

Newton's laws read:

Law I: Every body persists in its state of being at rest or of moving uniformly straight forward, except insofar as it is compelled to change its state by force impressed.

Law II: The alteration of motion is ever proportional to the motive force impress'd; and is made in the direction of the right line in which that force is impressed.

Law III: To every action there is always opposed an equal reaction: or the mutual actions of two bodies upon each other are always equal, and directed to contrary parts.

Newton's First Law

The first law states that if the net force (the vector sum of all forces acting on an object) is zero, then the velocity of the object is constant. Velocity is a vector quantity which expresses both the object's speed and the direction of its motion; therefore, the statement that the object's velocity is constant is a statement that both its speed and the direction of its motion are constant.

The first law can be stated mathematically when the mass is a non-zero constant, as,

$$\sum \mathbf{F} = 0 \Leftrightarrow \frac{d\mathbf{v}}{dt} = 0.$$

Consequently,

- An object that is at rest will stay at rest unless a force acts upon it.
- An object that is in motion will not change its velocity unless a force acts upon it.

This is known as *uniform motion*. An object *continues* to do whatever it happens to be doing unless a force is exerted upon it. If it is at rest, it continues in a state of rest (demonstrated when a table-cloth is skilfully whipped from under dishes on a tabletop and the dishes remain in their initial state of rest). If an object is moving, it continues to move without turning or changing its speed. This is evident in space probes that continuously move in outer space. Changes in motion must be imposed against the tendency of an object to retain its state of motion. In the absence of net forces, a moving object tends to move along a straight line path indefinitely.

Newton placed the first law of motion to establish frames of reference for which the other laws are applicable. However, Newton implicitly referred to the absolute co-ordinate of cosmos for this frame. Since we cannot precisely measure our velocity relative to a far star, Newton's frame is based on a pure imagination, not based on measurable physics. In current physics, an observer defines himself as in inertial frame by preparing one stone hooked by a spring, and rotating the spring to any direction, and observing the stone static and the length of that spring unchanged. By Einstein's equivalence principle, if there was one such observer A and another observer B moving in a constant velocity related to A, then A and B will both observe the same physics phenomena. if A verified the first law, then B will verify it too. In this way, the definition of inertial can get rid of absolute space or far star, and only refer to the objects locally reachable and measurable.

A particle not subject to forces moves (related to inertial frame) in a straight line at a constant speed. Newton's first law is often referred to as the *law of inertia*. Thus, a condition necessary for the uniform motion of a particle relative to an inertial reference frame is that the total net force acting on it is zero. In this sense, the first law can be restated as:

In every material universe, the motion of a particle in a preferential reference frame Φ is determined by the action of forces whose total vanished for all times when and only when the velocity of the particle is constant in Φ. That is, a particle initially at rest or in uniform motion in the preferential frame Φ continues in that state unless compelled by forces to change it.

Newton's first and second laws are valid only in an inertial reference frame. Any reference frame that is in uniform motion with respect to an inertial frame is also an inertial frame, i.e. Galilean invariance or the principle of Newtonian relativity.

Newton's Second Law

The second law states that the rate of change of momentum of a body is directly proportional to the force applied, and this change in momentum takes place in the direction of the applied force.

$$\mathbf{F} = \frac{d\mathbf{p}}{dt} = \frac{d(m\mathbf{v})}{dt}.$$

The second law can also be stated in terms of an object's acceleration. Since Newton's second law is valid only for constant-mass systems, m can be taken outside the differentiation operator by the constant factor rule in differentiation. Thus,

$$\mathbf{F} = m\frac{d\mathbf{v}}{dt} = m\mathbf{a},$$

where F is the net force applied, m is the mass of the body, and a is the body's acceleration. Thus, the net force applied to a body produces a proportional acceleration. In other words, if a body is accelerating, then there is a force on it. An application of this notation is the derivation of G Subscript C.

The above statements hint that the second law is merely a definition of **F**, not a precious observation of nature. However, current physics restate the second law in measurable steps: (1) defining the term 'one unit of mass' by a specified stone, (2) defining the term 'one unit of force' by a specified spring with specified length, (3) measuring by experiment or proving by theory (with a principle that every direction of space are equivalent), that force can be added as a mathematical vector, (4) finally conclude that $\mathbf{F} = m\mathbf{a}$. These steps hint the second law is a precious feature of nature.

The second law also implies the conservation of momentum: when the net force on the body is zero, the momentum of the body is constant. Any net force is equal to the rate of change of the momentum.

Any mass that is gained or lost by the system will cause a change in momentum that is not the result of an external force. A different equation is necessary for variable-mass systems.

Newton's second law is an approximation that is increasingly worse at high speeds because of relativistic effects.

According to modern ideas of how Newton was using his terminology, the law is understood, in modern terms, as an equivalent of:

> The change of momentum of a body is proportional to the impulse impressed on the body, and happens along the straight line on which that impulse is impressed.

This may be expressed by the formula F = p', where p' is the time derivative of the momentum p. This equation can be seen clearly in the Wren Library of Trinity College, Cambridge, in a glass case in which Newton's manuscript is open to the relevant page.

Motte's 1729 translation of Newton's Latin continued with Newton's commentary on the second law of motion, reading:

> If a force generates a motion, a double force will generate double the motion, a triple force triple the motion, whether that force be impressed altogether and at once, or gradually and successively. And this motion (being always directed the same way with the generating force), if the body moved before, is added to or subtracted from the former motion, according as they directly conspire with or are directly contrary to each other; or obliquely joined, when they are oblique, so as to produce a new motion compounded from the determination of both.

The sense or senses in which Newton used his terminology, and how he understood the second law and intended it to be understood, have been extensively discussed by historians of science, along with the relations between Newton's formulation and modern formulations.

Impulse

An impulse J occurs when a force F acts over an interval of time Δt, and it is given by:

$$\mathbf{J} = \int_{\Delta t} \mathbf{F}\, \mathrm{d}t.$$

Since force is the time derivative of momentum, it follows that:

$$\mathbf{J} = \Delta \mathbf{p} = m\Delta \mathbf{v}.$$

This relation between impulse and momentum is closer to Newton's wording of the second law.

Impulse is a concept frequently used in the analysis of collisions and impacts.

Variable-mass Systems

Variable-mass systems, like a rocket burning fuel and ejecting spent gases, are not closed and cannot be directly treated by making mass a function of time in the second law; that is, the following formula is wrong:

$$\mathbf{F}_{\mathrm{net}} = \frac{\mathrm{d}}{\mathrm{d}t}\big[m(t)\mathbf{v}(t)\big] = m(t)\frac{\mathrm{d}\mathbf{v}}{\mathrm{d}t} + \mathbf{v}(t)\frac{\mathrm{d}m}{\mathrm{d}t}. \qquad \text{(wrong)}$$

The falsehood of this formula can be seen by noting that it does not respect Galilean invariance: a variable-mass object with F = 0 in one frame will be seen to have F ≠ 0 in another frame. The correct equation of motion for a body whose mass m varies with time by either ejecting or accreting mass is obtained by applying the second law to the entire, constant-mass system consisting of the body and its ejected/accreted mass; the result is:

$$\mathbf{F} + \mathbf{u}\frac{dm}{dt} = m\frac{d\mathbf{v}}{dt}$$

where u is the velocity of the escaping or incoming mass relative to the body. From this equation one can derive the equation of motion for a varying mass system, for example, the Tsiolkovsky rocket equation. Under some conventions, the quantity $u\,dm/dt$ on the left-hand side, which represents the advection of momentum, is defined as a force (the force exerted on the body by the changing mass, such as rocket exhaust) and is included in the quantity F. Then, by substituting the definition of acceleration, the equation becomes F = ma.

Newton's Third Law

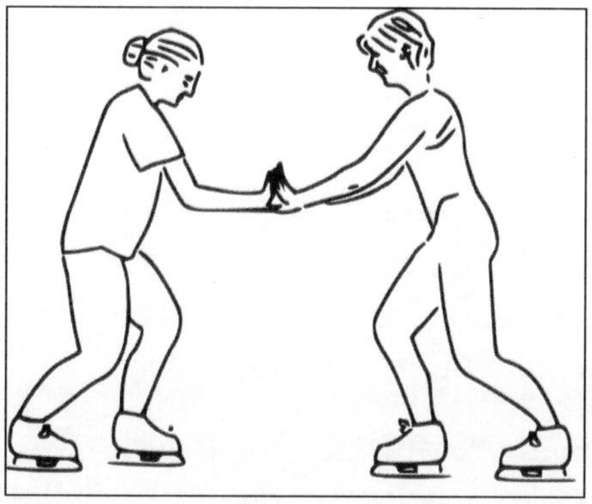

An illustration of Newton's third law in which two skaters push against each other. The first skater on the left exerts a normal force N_{12} on the second skater directed towards the right, and the second skater exerts a normal force N_{21} on the first skater directed towards the left. The magnitudes of both forces are equal, but they have opposite directions, as dictated by Newton's third law.

The third law states that all forces between two objects exist in equal magnitude and opposite direction: if one object A exerts a force FA on a second object B, then B simultaneously exerts a force FB on A, and the two forces are equal in magnitude and opposite in direction: FA = −FB. The third law means that all forces are interactions between different bodies, or different regions within one body, and thus that there is no such thing as a force that is not accompanied by an equal and opposite force. In some situations, the magnitude and direction of the forces are determined entirely by one of the two bodies, say Body A; the force exerted by Body A on Body B is called the "action", and the force exerted by Body B on Body A is called the "reaction". This law is sometimes referred to as the action-reaction law, with FA called the "action" and FB the "reaction". In other situations the magnitude and directions of the forces are determined jointly by both bodies and it isn't necessary to identify one force as the "action" and the other as the "reaction". The action and the reaction are simultaneous, and it does not matter which is called the action and which is called reaction; both forces are part of a single interaction, and neither force exists without the other.

The two forces in Newton's third law are of the same type (e.g., if the road exerts a forward frictional force on an accelerating car's tires, then it is also a frictional force that Newton's third law predicts for the tires pushing backward on the road).

From a conceptual standpoint, Newton's third law is seen when a person walks: they push against the floor, and the floor pushes against the person. Similarly, the tires of a car push against the road while the road pushes back on the tires—the tires and road simultaneously push against each other. In swimming, a person interacts with the water, pushing the water backward, while the water simultaneously pushes the person forward—both the person and the water push against each other. The reaction forces account for the motion in these examples. These forces depend on friction; a person or car on ice, for example, may be unable to exert the action force to produce the needed reaction force.

Newton used the third law to derive the law of conservation of momentum; from a deeper perspective, however, conservation of momentum is the more fundamental idea (derived via Noether's theorem from Galilean invariance), and holds in cases where Newton's third law appears to fail, for instance when force fields as well as particles carry momentum, and in quantum mechanics.

Importance and Range of Validity

Newton's laws were verified by experiment and observation for over 200 years, and they are excellent approximations at the scales and speeds of everyday life. Newton's laws of motion, together with his law of universal gravitation and the mathematical techniques of calculus, provided for the first time a unified quantitative explanation for a wide range of physical phenomena.

These three laws hold to a good approximation for macroscopic objects under everyday conditions. However, Newton's laws (combined with universal gravitation and classical electrodynamics) are inappropriate for use in certain circumstances, most notably at very small scales, at very high speeds, or in very strong gravitational fields. Therefore, the laws cannot be used to explain phenomena such as conduction of electricity in a semiconductor, optical properties of substances, errors in non-relativistically corrected GPS systems and superconductivity. Explanation of these phenomena requires more sophisticated physical theories, including general relativity and quantum field theory.

In quantum mechanics, concepts such as force, momentum, and position are defined by linear operators that operate on the quantum state; at speeds that are much lower than the speed of light, Newton's laws are just as exact for these operators as they are for classical objects. At speeds comparable to the speed of light, the second law holds in the original form $F = dp/dt$, where F and p are four-vectors.

Relationship to the Conservation Laws

In modern physics, the laws of conservation of momentum, energy, and angular momentum are of more general validity than Newton's laws, since they apply to both light and matter, and to both classical and non-classical physics.

This can be stated simply, "Momentum, energy and angular momentum cannot be created or destroyed".

Because force is the time derivative of momentum, the concept of force is redundant and subordinate to the conservation of momentum, and is not used in fundamental theories (e.g., quantum mechanics, quantum electrodynamics, general relativity, etc.). The standard model explains in detail how the three fundamental forces known as gauge forces originate out of exchange by virtual particles. Other forces, such as gravity and fermionic degeneracy pressure, also arise from the momentum conservation. Indeed, the conservation of 4-momentum in inertial motion via curved space-time results in what we call gravitational force in general relativity theory. The application of the space derivative (which is a momentum operator in quantum mechanics) to the overlapping wave functions of a pair of fermions (particles with half-integer spin) results in shifts of maxima of compound wavefunction away from each other, which is observable as the "repulsion" of the fermions.

Newton stated the third law within a world-view that assumed instantaneous action at a distance between material particles. However, he was prepared for philosophical criticism of this action at a distance, and it was in this context that he stated the famous phrase, "I feign no hypotheses". In modern physics, action at a distance has been completely eliminated, except for subtle effects involving quantum entanglement. (In particular, this refers to Bell's theorem—that no local model can reproduce the predictions of quantum theory.) Despite only being an approximation, in modern engineering and all practical applications involving the motion of vehicles and satellites, the concept of action at a distance is used extensively.

The discovery of the second law of thermodynamics by Carnot in the 19th century showed that not every physical quantity is conserved over time, thus disproving the validity of inducing the opposite metaphysical view from Newton's laws. Hence, a "steady-state" worldview based solely on Newton's laws and the conservation laws does not take entropy into account.

Friction

Friction is the force resisting the relative motion of solid surfaces, fluid layers, and material elements sliding against each other. There are several types of friction:

- Dry friction is a force that opposes the relative lateral motion of two solid surfaces in contact. Dry friction is subdivided into static friction ("stiction") between non-moving surfaces, and kinetic friction between moving surfaces. With the exception of atomic or molecular friction, dry friction generally arises from the interaction of surface features, known as asperities.

- Fluid friction describes the friction between layers of a viscous fluid that are moving relative to each other.

- Lubricated friction is a case of fluid friction where a lubricant fluid separates two solid surfaces.

- Skin friction is a component of drag, the force resisting the motion of a fluid across the surface of a body.

- Internal friction is the force resisting motion between the elements making up a solid material while it undergoes deformation.

When surfaces in contact move relative to each other, the friction between the two surfaces converts kinetic energy into thermal energy (that is, it converts work to heat). This property can have dramatic consequences, as illustrated by the use of friction created by rubbing pieces of wood together to start a fire. Kinetic energy is converted to thermal energy whenever motion with friction occurs, for example when a viscous fluid is stirred. Another important consequence of many types of friction can be wear, which may lead to performance degradation or damage to components. Friction is a component of the science of tribology.

Friction is desirable and important in supplying traction to facilitate motion on land. Most land vehicles rely on friction for acceleration, deceleration and changing direction. Sudden reductions in traction can cause loss of control and accidents.

Friction is not itself a fundamental force. Dry friction arises from a combination of inter-surface adhesion, surface roughness, surface deformation, and surface contamination. The complexity of these interactions makes the calculation of friction from first principles impractical and necessitates the use of empirical methods for analysis and the development of theory.

Friction is a non-conservative force - work done against friction is path dependent. In the presence of friction, some kinetic energy is always transformed to thermal energy, so mechanical energy is not conserved.

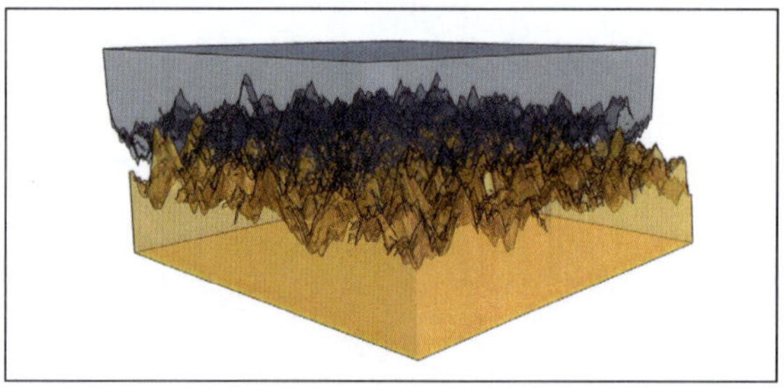

Simulated blocks with fractal rough surfaces, exhibiting static frictional interactions.

Laws of Dry Friction

The elementary property of sliding (kinetic) friction were discovered by experiment in the 15th to 18th centuries and were expressed as three empirical laws:

- Amontons' First Law: The force of friction is directly proportional to the applied load.
- Amontons' Second Law: The force of friction is independent of the apparent area of contact.
- Coulomb's Law of Friction: Kinetic friction is independent of the sliding velocity.

Dry Friction

Dry friction resists relative lateral motion of two solid surfaces in contact. The two regimes of dry friction are 'static friction' ("stiction") between non-moving surfaces, and kinetic friction (sometimes called sliding friction or dynamic friction) between moving surfaces.

Coulomb friction, named after Charles-Augustin de Coulomb, is an approximate model used to calculate the force of dry friction. It is governed by the model:

$$F_f \leq \mu F_n,$$

where,

- F_f is the force of friction exerted by each surface on the other. It is parallel to the surface, in a direction opposite to the net applied force.
- μ is the coefficient of friction, which is an empirical property of the contacting materials.
- F_n is the normal force exerted by each surface on the other, directed perpendicular (normal) to the surface.

The Coulomb friction F_f may take any value from zero up to μF_n, and the direction of the frictional force against a surface is opposite to the motion that surface would experience in the absence of friction. Thus, in the static case, the frictional force is exactly what it must be in order to prevent motion between the surfaces; it balances the net force tending to cause such motion. In this case, rather than providing an estimate of the actual frictional force, the Coulomb approximation provides a threshold value for this force, above which motion would commence. This maximum force is known as traction.

The force of friction is always exerted in a direction that opposes movement (for kinetic friction) or potential movement (for static friction) between the two surfaces. For example, a curling stone sliding along the ice experiences a kinetic force slowing it down. For an example of potential movement, the drive wheels of an accelerating car experience a frictional force pointing forward; if they did not, the wheels would spin, and the rubber would slide backwards along the pavement. Note that it is not the direction of movement of the vehicle they oppose, it is the direction of (potential) sliding between tire and road.

Normal Force

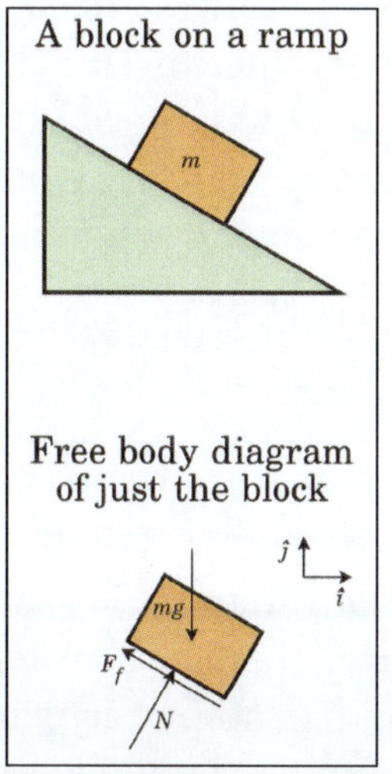

Free-body diagram for a block on a ramp. Arrows are vectors indicating directions and magnitudes of forces. *N* is the normal force, *mg* is the force of gravity, and *Ff* is the force of friction.

The normal force is defined as the net force compressing two parallel surfaces together, and its direction is perpendicular to the surfaces. In the simple case of a mass resting on a horizontal surface, the only component of the normal force is the force due to gravity, where $N = mg$. In this case, the magnitude of the friction force is the product of the mass of the object, the acceleration due to gravity, and the coefficient of friction. However, the coefficient of friction is not a function of mass or volume; it depends only on the material. For instance, a large aluminum block has the same coefficient of friction as a small aluminum block. However, the magnitude of the friction force itself depends on the normal force, and hence on the mass of the block.

If an object is on a level surface and the force tending to cause it to slide is horizontal, the normal force N between the object and the surface is just its weight, which is equal to its mass multiplied by the acceleration due to earth's gravity, g. If the object is on a tilted surface such as an inclined plane, the normal force is less, because less of the force of gravity is perpendicular to the face of the plane. Therefore, the normal force, and ultimately the frictional force, is determined using vector analysis, usually via a free body diagram. Depending on the situation, the calculation of the normal force may include forces other than gravity.

Coefficient of Friction

The coefficient of friction (COF), often symbolized by the Greek letter μ, is a dimensionless scalar value which describes the ratio of the force of friction between two bodies and the force pressing them together. The coefficient of friction depends on the materials used; for example, ice on steel has a low coefficient of friction, while rubber on pavement has a high coefficient of friction. Coefficients of friction range from near zero to greater than one. It is an axiom of the nature of friction between metal surfaces that it is greater between two surfaces of similar metals than between two surfaces of different metals— hence, brass will have a higher coefficient of friction when moved against brass, but less if moved against steel or aluminum.

For surfaces at rest relative to each other $\mu = \mu_s$, where μ_s is the coefficient of static friction. This is usually larger than its kinetic counterpart. The coefficient of static friction exhibited by a pair of contacting surfaces depends upon the combined effects of material deformation characteristics and surface roughness, both of which have their origins in the chemical bonding between atoms in each of the bulk materials and between the material surfaces and any adsorbed material. The fractality of surfaces, a parameter describing the scaling behavior of surface asperities, is known to play an important role in determining the magnitude of the static friction.

For surfaces in relative motion $\mu = \mu_k$, where μ_k is the coefficient of kinetic friction. The Coulomb friction is equal to F_f, and the frictional force on each surface is exerted in the direction opposite to its motion relative to the other surface.

Arthur Morin introduced the term and demonstrated the utility of the coefficient of friction. The coefficient of friction is an empirical measurement – it has to be measured experimentally, and cannot be found through calculations. Rougher surfaces tend to have higher effective values. Both static and kinetic coefficients of friction depend on the pair of surfaces in contact; for a given pair of surfaces, the coefficient of static friction is *usually* larger than that of kinetic friction; in some sets the two coefficients are equal, such as teflon-on-teflon.

Most dry materials in combination have friction coefficient values between 0.3 and 0.6. Values outside this range are rarer, but teflon, for example, can have a coefficient as low as 0.04. A value of zero would mean no friction at all, an elusive property. Rubber in contact with other surfaces can yield friction coefficients from 1 to 2. Occasionally it is maintained that μ is always < 1, but this is not true. While in most relevant applications μ < 1, a value above 1 merely implies that the force required to slide an object along the surface is greater than the normal force of the surface on the object. For example, silicone rubber or acrylic rubber-coated surfaces have a coefficient of friction that can be substantially larger than 1.

While it is often stated that the COF is a "material property," it is better categorized as a "system property." Unlike true material properties (such as conductivity, dielectric constant, yield strength), the COF for any two materials depends on system variables like temperature, velocity, atmosphere and also what are now popularly described as aging and deaging times; as well as on geometric properties of the interface between the materials, namely surface structure. For example, a copper pin sliding against a thick copper plate can have a COF that varies from 0.6 at low speeds (metal sliding against metal) to below 0.2 at high speeds when the copper surface begins to melt due to frictional heating. The latter speed, of course, does not determine the COF uniquely; if the pin diameter is increased so that the frictional heating is removed rapidly, the temperature drops, the pin remains solid and the COF rises to that of a 'low speed' test.

Approximate Coefficients of Friction

Materials		Static Friction,		Kinetic/Sliding Friction,	
Dry and clean		Lubricated	Dry and clean	Lubricated	
Aluminium	Steel	0.61		0.47[	
Aluminum	Aluminum	1.05-1.35	0.3	1.4-1.5	
Gold	Gold			2.5	
Platinum	Platinum	1.2[	0.25	3.0	
Silver	Silver	1.4	0.55	1.5	
Alumina ceramic	Silicon Nitride ceramic				0.004 (wet)
BAM (Ceramic alloy $AlMgB_{14}$)	Titanium boride (TiB_2)	0.04–0.05	0.02		
Brass	Steel	0.35-0.51	0.19	0.44	
Cast iron	Copper	1.05		0.29	
Cast iron	Zinc	0.85		0.21	
Concrete	Rubber	1.0	0.30 (wet)	0.6-0.85	0.45-0.75 (wet)
Concrete	Wood	0.62			
Copper	Glass	0.68			
Copper	Steel	0.53		0.36	
Glass	Glass	0.9-1.0		0.4	
Human synovial fluid	Cartilage		0.01		0.003

Ice	Ice	0.02-0.09			
Polyethene	Steel	0.2	0.2		
PTFE (Teflon)	PTFE (Teflon)	0.04	0.04		0.04
Steel	Ice	0.03			
Steel	PTFE (Teflon)	0.04-0.2	0.04		0.04
Steel	Steel	0.740.80	0.16	0.42-0.62	
Wood	Metal	0.2–0.6	0.2 (wet)		
Wood	Wood	0.25–0.5	0.2 (wet)		

Under certain conditions some materials have very low friction coefficients. An example is (highly ordered pyrolytic) graphite which can have a friction coefficient below 0.01. This ultralow-friction regime is called superlubricity.

Static Friction

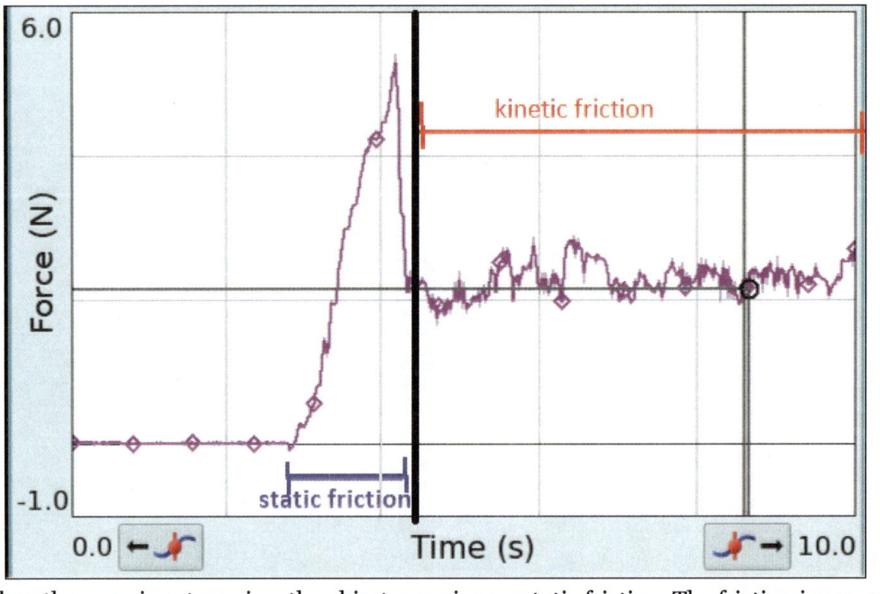

When the mass is not moving, the object experiences static friction. The friction increases as the applied force increases until the block moves. After the block moves, it experiences kinetic friction, which is less than the maximum static friction.

Static friction is friction between two or more solid objects that are not moving relative to each other. For example, static friction can prevent an object from sliding down a sloped surface. The coefficient of static friction, typically denoted as μ_s, is usually higher than the coefficient of kinetic friction. Static friction is considered to arise as the result of surface roughness features across multiple length-scales at solid surfaces. These features, known as asperities are present down to nano-scale dimensions and result in true solid to solid contact existing only at a limited number of points accounting for only a fraction of the apparent or nominal contact area . The linearity between applied load and true contact area, arising from asperity deformation, gives rise to the linearity between static frictional force and normal force, found for typical Amonton-Coulomb type friction.

The static friction force must be overcome by an applied force before an object can move. The maximum possible friction force between two surfaces before sliding begins is the product of the coefficient of static friction and the normal force: $F_{max} = \mu_s F_n$. When there is no sliding occurring, the friction force can have any value from zero up to $_{max}$. Any force smaller than F_{max} attempting to slide one surface over the other is opposed by a frictional force of equal magnitude and opposite direction. Any force larger than F_{max} overcomes the force of static friction and causes sliding to occur. The instant sliding occurs, static friction is no longer applicable—the friction between the two surfaces is then called kinetic friction.

An example of static friction is the force that prevents a car wheel from slipping as it rolls on the ground. Even though the wheel is in motion, the patch of the tire in contact with the ground is stationary relative to the ground, so it is static rather than kinetic friction.

The maximum value of static friction, when motion is impending, is sometimes referred to as limiting friction, although this term is not used universally.

Kinetic Friction

Kinetic friction, also known as dynamic friction or sliding friction, occurs when two objects are moving relative to each other and rub together (like a sled on the ground). The coefficient of kinetic friction is typically denoted as μ_k, and is usually less than the coefficient of static friction for the same materials. However, Richard Feynman comments that "with dry metals it is very hard to show any difference." The friction force between two surfaces after sliding begins is the product of the coefficient of kinetic friction and the normal force:

$$F_k = \mu_k F_n.$$

New models are beginning to show how kinetic friction can be greater than static friction. Kinetic friction is now understood, in many cases, to be primarily caused by chemical bonding between the surfaces, rather than interlocking asperities; however, in many other cases roughness effects are dominant, for example in rubber to road friction. Surface roughness and contact area affect kinetic friction for micro- and nano-scale objects where surface area forces dominate inertial forces.

The origin of kinetic friction at nanoscale can be explained by thermodynamics. Upon sliding, new surface forms at the back of a sliding true contact, and existing surface disappears at the front of it. Since all surfaces involve the thermodynamic surface energy, work must be spent in creating the new surface, and energy is released as heat in removing the surface. Thus, a force is required to move the back of the contact, and frictional heat is released at the front.

Angle of Friction

For certain applications, it is more useful to define static friction in terms of the maximum angle before which one of the items will begin sliding. This is called the *angle of friction* or *friction angle*. It is defined as:

$$\tan\theta = \mu_s$$

where θ is the angle from horizontal and μs is the static coefficient of friction between the objects.

This formula can also be used to calculate μs from empirical measurements of the friction angle.

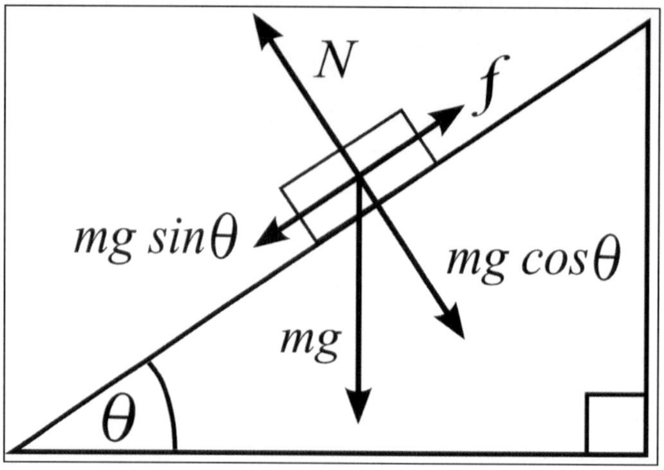

Angle of friction, θ, when block just starts to slide.

Friction at the Atomic Level

Determining the forces required to move atoms past each other is a challenge in designing nanomachines. In 2008 scientists for the first time were able to move a single atom across a surface, and measure the forces required. Using ultrahigh vacuum and nearly zero temperature (5° K), a modified atomic force microscope was used to drag a cobalt atom, and a carbon monoxide molecule, across surfaces of copper and platinum.

Limitations of the Coulomb Model

The Coulomb approximation follows from the assumptions that: surfaces are in atomically close contact only over a small fraction of their overall area; that this contact area is proportional to the normal force (until saturation, which takes place when all area is in atomic contact); and that the frictional force is proportional to the applied normal force, independently of the contact area. The Coulomb approximation is fundamentally an empirical construct. It is a rule-of-thumb describing the approximate outcome of an extremely complicated physical interaction. The strength of the approximation is its simplicity and versatility. Though the relationship between normal force and frictional force is not exactly linear (and so the frictional force is not entirely independent of the contact area of the surfaces), the Coulomb approximation is an adequate representation of friction for the analysis of many physical systems.

When the surfaces are conjoined, Coulomb friction becomes a very poor approximation (for example, adhesive tape resists sliding even when there is no normal force, or a negative normal force). In this case, the frictional force may depend strongly on the area of contact. Some drag racing tires are adhesive for this reason. However, despite the complexity of the fundamental physics behind friction, the relationships are accurate enough to be useful in many applications.

Negative Coefficient of Friction

As of 2012, a single study has demonstrated the potential for an effectively negative coefficient of friction in the low-load regime, meaning that a decrease in normal force leads to an increase in friction. This contradicts everyday experience in which an increase in normal force leads to an increase in friction. This was reported in the journal *Nature* in October 2012 and involved the friction encountered by an atomic force microscope stylus when dragged across a graphene sheet in the presence of graphene-adsorbed oxygen.

Numerical Simulation of the Coulomb Model

Despite being a simplified model of friction, the Coulomb model is useful in many numerical simulation applications such as multibody systems and granular material. Even its most simple expression encapsulates the fundamental effects of sticking and sliding which are required in many applied cases, although specific algorithms have to be designed in order to efficiently numerically integrate mechanical systems with Coulomb friction and bilateral or unilateral contact. Some quite nonlinear effects, such as the so-called Painlevé paradoxes, may be encountered with Coulomb friction.

Dry Friction and Instabilities

Dry friction can induce several types of instabilities in mechanical systems which display a stable behaviour in the absence of friction. These instabilities may be caused by the decrease of the friction force with an increasing velocity of sliding, by material expansion due to heat generation during friction (the thermo-elastic instabilities), or by pure dynamic effects of sliding of two elastic materials (the Adams-Martins instabilities). The latter were originally discovered in 1995 by George G. Adams and João Arménio Correia Martins for smooth surfaces and were later found in periodic rough surfaces. In particular, friction-related dynamical instabilities are thought to be responsible for brake squeal and the 'song' of a glass harp, phenomena which involve stick and slip, modelled as a drop of friction coefficient with velocity.

A practically important case is the self-oscillation of the strings of bowed instruments such as the violin, cello, hurdy-gurdy, erhu, etc.

A connection between dry friction and flutter instability in a simple mechanical system has been discovered.

Frictional instabilities can lead to the formation of new self-organized patterns (or "secondary structures") at the sliding interface, such as in-situ formed tribofilms which are utilized for the reduction of friction and wear in so-called self-lubricating materials.

Fluid Friction

Fluid friction occurs between fluid layers that are moving relative to each other. This internal resistance to flow is named *viscosity*. In everyday terms, the viscosity of a fluid is described as its "thickness". Thus, water is "thin", having a lower viscosity, while honey is "thick", having a higher viscosity. The less viscous the fluid, the greater its ease of deformation or movement.

All real fluids (except superfluids) offer some resistance to shearing and therefore are viscous. For teaching and explanatory purposes it is helpful to use the concept of an inviscid fluid or an ideal fluid which offers no resistance to shearing and so is not viscous.

Lubricated Friction

Lubricated friction is a case of fluid friction where a fluid separates two solid surfaces. Lubrication is a technique employed to reduce wear of one or both surfaces in close proximity moving relative to each another by interposing a substance called a lubricant between the surfaces.

In most cases the applied load is carried by pressure generated within the fluid due to the frictional viscous resistance to motion of the lubricating fluid between the surfaces. Adequate lubrication allows smooth continuous operation of equipment, with only mild wear, and without excessive stresses or seizures at bearings. When lubrication breaks down, metal or other components can rub destructively over each other, causing heat and possibly damage or failure.

Skin Friction

Skin friction arises from the interaction between the fluid and the skin of the body, and is directly related to the area of the surface of the body that is in contact with the fluid. Skin friction follows the drag equation and rises with the square of the velocity.

Skin friction is caused by viscous drag in the boundary layer around the object. There are two ways to decrease skin friction: the first is to shape the moving body so that smooth flow is possible, like an airfoil. The second method is to decrease the length and cross-section of the moving object as much as is practicable.

Internal Friction

Internal friction is the force resisting motion between the elements making up a solid material while it undergoes deformation.

Plastic deformation in solids is an irreversible change in the internal molecular structure of an object. This change may be due to either (or both) an applied force or a change in temperature. The change of an object's shape is called strain. The force causing it is called stress.

Elastic deformation in solids is reversible change in the internal molecular structure of an object. Stress does not necessarily cause permanent change. As deformation occurs, internal forces oppose the applied force. If the applied stress is not too large these opposing forces may completely resist the applied force, allowing the object to assume a new equilibrium state and to return to its original shape when the force is removed. This is known as elastic deformation or elasticity.

Radiation Friction

As a consequence of light pressure, Einstein in 1909 predicted the existence of "radiation friction" which would oppose the movement of matter. He wrote, "radiation will exert pressure on both sides of the plate. The forces of pressure exerted on the two sides are equal if the plate is at rest. However, if it is in motion, more radiation will be reflected on the surface that is ahead during the motion (front surface) than on the back surface. The backwardacting force of pressure exerted on the front surface is thus larger than the force of pressure acting on the back. Hence, as the resultant of the two forces, there remains a force that counteracts the motion of the plate and that increases with the velocity of the plate. We will call this resultant ‘radiation friction’ in brief”.

Other Types of Friction

Rolling Resistance

Rolling resistance is the force that resists the rolling of a wheel or other circular object along a surface caused by deformations in the object or surface. Generally the force of rolling resistance is less than that associated with kinetic friction. Typical values for the coefficient of rolling resistance are 0.001. One of the most common examples of rolling resistance is the movement of motor vehicle tires on a road, a process which generates heat and sound as by-products.

Braking Friction

Any wheel equipped with a brake is capable of generating a large retarding force, usually for the purpose of slowing and stopping a vehicle or piece of rotating machinery. Braking friction differs from rolling friction because the coefficient of friction for rolling friction is small whereas the coefficient of friction for braking friction is designed to be large by choice of materials for brake pads.

Triboelectric Effect

Rubbing dissimilar materials against one another can cause a build-up of electrostatic charge, which can be hazardous if flammable gases or vapours are present. When the static build-up discharges, explosions can be caused by ignition of the flammable mixture.

Belt Friction

Belt friction is a physical property observed from the forces acting on a belt wrapped around a pulley, when one end is being pulled. The resulting tension, which acts on both ends of the belt, can be modeled by the belt friction equation.

In practice, the theoretical tension acting on the belt or rope calculated by the belt friction equation can be compared to the maximum tension the belt can support. This helps a designer of such a rig to know how many times the belt or rope must be wrapped around the pulley to prevent it from slipping. Mountain climbers and sailing crews demonstrate a standard knowledge of belt friction when accomplishing basic tasks.

Reducing Friction

Devices

Devices such as wheels, ball bearings, roller bearings, and air cushion or other types of fluid bearings can change sliding friction into a much smaller type of rolling friction.

Many thermoplastic materials such as nylon, HDPE and PTFE are commonly used in low friction bearings. They are especially useful because the coefficient of friction falls with increasing imposed load. For improved wear resistance, very high molecular weight grades are usually specified for heavy duty or critical bearings.

Lubricants

A common way to reduce friction is by using a lubricant, such as oil, water, or grease, which is placed between the two surfaces, often dramatically lessening the coefficient of friction. The science of friction and lubrication is called tribology. Lubricant technology is when lubricants are mixed with the application of science, especially to industrial or commercial objectives.

Superlubricity, a recently discovered effect, has been observed in graphite: it is the substantial decrease of friction between two sliding objects, approaching zero levels. A very small amount of frictional energy would still be dissipated.

Lubricants to overcome friction need not always be thin, turbulent fluids or powdery solids such as graphite and talc; acoustic lubrication actually uses sound as a lubricant.

Another way to reduce friction between two parts is to superimpose micro-scale vibration to one of the parts. This can be sinusoidal vibration as used in ultrasound-assisted cutting or vibration noise, known as dither.

Energy of Friction

According to the law of conservation of energy, no energy is destroyed due to friction, though it may be lost to the system of concern. Energy is transformed from other forms into thermal energy. A sliding hockey puck comes to rest because friction converts its kinetic energy into heat which raises the thermal energy of the puck and the ice surface. Since heat quickly dissipates, many early philosophers, including Aristotle, wrongly concluded that moving objects lose energy without a driving force.

When an object is pushed along a surface along a path C, the energy converted to heat is given by a line integral, in accordance with the definition of work:

$$E_{th} = \int_C F_{fric}(\mathbf{x}) \cdot d\mathbf{x} = \int_C \mu_k \, \mathbf{F}_n(\mathbf{x}) \cdot d\mathbf{x},$$

where,

$\mathbf{F}_{fric}$ is the friction force,

$\mathbf{F}_n$ is the vector obtained by multiplying the magnitude of the normal force by a unit vector pointing *against* the object's motion,

μ_k is the coefficient of kinetic friction, which is inside the integral because it may vary from location to location (e.g. if the material changes along the path),

x is the position of the object.

Energy lost to a system as a result of friction is a classic example of thermodynamic irreversibility.

Work of Friction

In the reference frame of the interface between two surfaces, static friction does *no* work, because there is never displacement between the surfaces. In the same reference frame, kinetic friction is always in the direction opposite the motion, and does *negative* work. However, friction can do *positive* work in certain frames of reference. One can see this by placing a heavy box on a rug, then pulling on the rug quickly. In this case, the box slides backwards relative to the rug, but moves forward relative to the frame of reference in which the floor is stationary. Thus, the kinetic friction between the box and rug accelerates the box in the same direction that the box moves, doing *positive* work.

The work done by friction can translate into deformation, wear, and heat that can affect the contact surface properties (even the coefficient of friction between the surfaces). This can be beneficial as in polishing. The work of friction is used to mix and join materials such as in the process of friction welding. Excessive erosion or wear of mating sliding surfaces occurs when work due to frictional forces rise to unacceptable levels. Harder corrosion particles caught between mating surfaces in relative motion (fretting) exacerbates wear of frictional forces. Bearing seizure or failure may result from excessive wear due to work of friction. As surfaces are worn by work due to friction, fit and surface finish of an object may degrade until it no longer functions properly.

Applications

Friction is an important factor in many engineering disciplines.

Transportation

- Automobile brakes inherently rely on friction, slowing a vehicle by converting its kinetic energy into heat. Incidentally, dispersing this large amount of heat safely is one technical challenge in designing brake systems. Disk brakes rely on friction between a disc and brake pads that are squeezed transversely against the rotating disc. In drum brakes, brake shoes or pads are pressed outwards against a rotating cylinder (brake drum) to create friction. Since braking discs can be more efficiently cooled than drums, disc brakes have better stopping performance.

- Rail adhesion refers to the grip wheels of a train have on the rails, see Frictional contact mechanics.

- Road slipperiness is an important design and safety factor for automobiles:

    - Split friction is a particularly dangerous condition arising due to varying friction on either side of a car.

    - Road texture affects the interaction of tires and the driving surface.

Measurement

- A tribometer is an instrument that measures friction on a surface.

- A profilograph is a device used to measure pavement surface roughness.

Household Usage

- Friction is used to heat and ignite matchsticks (friction between the head of a matchstick and the rubbing surface of the match box).
- Sticky pads are used to prevent object from slipping off smooth surfaces by effectively increasing the friction coefficient between the surface and the object.

Law of Restitution

Newton's law of restitution says that when two objects collide, their speeds after the collision depend on the material from which they are made.

This measure of the nature of the materials colliding is represented by a number called the coefficient of restitution. This coefficient is equal to the relative speed between the two objects after the collision, divided by the relative speed between the two objects before the collision:

$$e = \frac{\text{relative speed after collision}}{\text{relative speed before collision}}$$

This is a useful measure, because it tells you something about the 'bounciness,' or elasticity, of the collision. A collision where no overall kinetic (movement) energy is lost is called a perfectly elastic collision, and these are the most bouncy. They have a coefficient of restitution of $e = 1$. A collision where the maximum kinetic (movement) energy is lost is called a perfectly inelastic collision, and this is where the two objects stick together and move together after the collision. They have a coefficient of restitution of $e = 0$. And a number in between is in between. It's an inelastic collision where some kinetic energy is lost. Most real-life collisions are in between.

With this equation, you always take the smaller number away from the larger, so it never comes out as a negative number. The coefficient of restitution is always positive.

Collision between Two Moving Objects

When two objects collide, we can use the equation previously mentioned to calculate the coefficient of restitution. Or if you're given the coefficient of restitution, you can use it to find a missing velocity.

For example, let's say one car rear-ends another on a nearby highway. Car A was going at 70 miles per hour, and car B was going at 50 miles per hour. After the collision, Car A is going 40 miles per hour in the same direction, and car B is going 60 miles per hour. What is the coefficient of restitution between the two cars?

Well, we need to figure out the relative speed between the two cars. That's just the difference of the two velocities. Knowing that leads to an expanded equation: The two cars are going in the same direction, so both have positive velocities. Let's plug the numbers into the equation:

$$e = \frac{V_{fb} - V_{fa}}{V_{ia} - V_{ib}}$$

V_{fa} = final velocity of A

V_{fb} = final velocity of B

V_{ia} = initial velocity of A

V_{ib} = initial velocity of B

$$e = \frac{V_{fb} - V_{fa}}{V_{ia} - V_{ib}}$$

$$e = \frac{60 - 40}{70 - 50} = \frac{20}{20} = 1$$

So the coefficient of restitution is 1. That means that it is an elastic collision.

Principle of Transmissibility

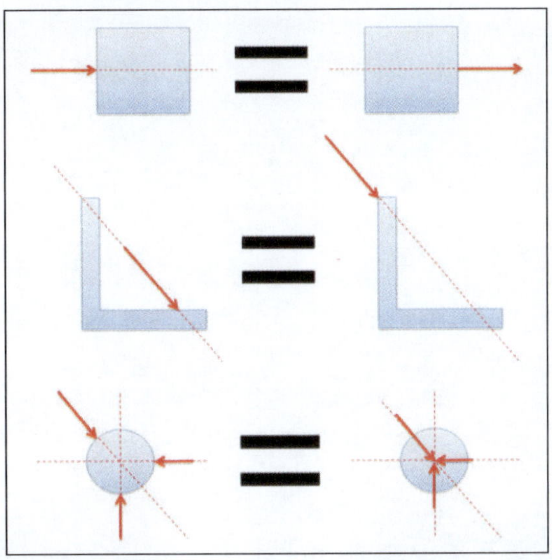

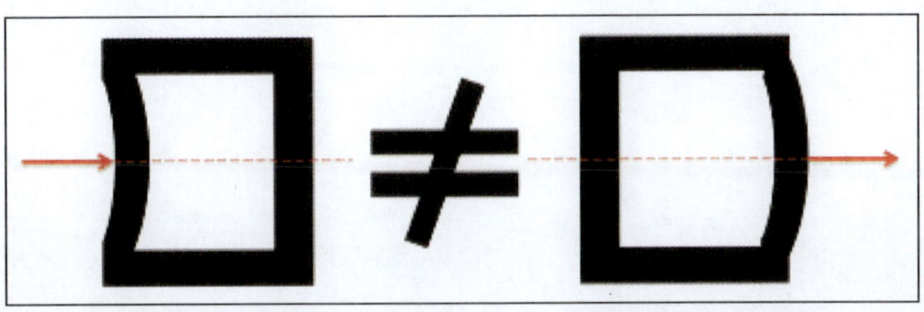

The exact point of application of a force will impact how internal forces (stresses) are distributed, so the principle of transmissibility cannot be applied when examining internal forces.

The principle of transmissibility states that the point of application of a force can be moved anywhere along its line of action without changing the external reaction forces on a rigid body. Any force that has the same magnitude and direction, and which has a point of application somewhere along the same line of action will cause the same acceleration and will result in the same moment. Therefore, the points of application of forces may be moved along the line of action to simplify the analysis of rigid bodies.

References

- Fundamental-concepts-principle-applied-mechanics: polytechnichub.com, Retrieved 16 April, 2019
- Vectors, science-mechanics: britannica.com, Retrieved 28 March, 2019
- Z.Y.Wang (2016). "Generalized momentum equation of quantum mechanics". Optical and Quantum Electronics. 48 (2): 1–9. doi:10.1007/s11082-015-0261-8
- Hanaor, D.; Gan, Y.; Einav, I. (2016). "Static friction at fractal interfaces". Tribology International. 93: 229–238. doi:10.1016/j.triboint.2015.09.016
- Newton-s-law-of-restitution-definition-use: study.com, Retrieved 15 June, 2019
- Principle-of-transmissibility, 2-equilibrium-concurrent: adaptivemap.ma.psu.edu, Retrieved 10 January, 2019

Statics and Force 3

- **Statics**
- **Force**
- **Force Systems**
- **Moment of Force**
- **Moment of Couples**

Statics comes under the branch of mechanics that deals with the analysis of force and torque of physical systems in static equilibrium with their environment. Force is the resultant action of push or pull on interaction between objects. The topics elaborated in this chapter will provide an easy understanding of the concepts related to statics and force.

Statics

Statics is the subdivision of mechanics that is concerned with the forces that act on bodies at rest under equilibrium conditions. Its foundations were laid more than 2,200 years ago by the ancient Greek mathematician Archimedes and others while studying the force-amplifying properties of simple machines such as the lever and the axle. The methods and results of the science of statics have proved especially useful in designing buildings, bridges, and dams, as well as cranes and other similar mechanical devices. To be able to calculate the dimensions of such structures and machines, architects and engineers must first determine the forces that act on their interconnected parts. Statics provides the analytical and graphical procedures needed to identify and describe these unknown forces.

Statics assumes that the bodies with which it deals are perfectly rigid. It also holds that the sum of all the forces acting on a body at rest has to be zero (i.e., the forces involved balance one another) and that there must be no tendency for the forces to turn the body about any axis. These three conditions are independent of one another, and their expression in mathematical form comprises the equations of equilibrium. There are three equations, and so only three unknown forces can be calculated. If more than three unknown forces exist, it means that there are more components in the structure or machine than are required to support the applied loads or that there are more restraints than are needed to keep the body from moving. Such unnecessary components or restraints are termed redundant (e.g., a table with four legs has one redundant leg) and the system of forces is said to be statically indeterminate. The number of equations available in statics is limited because of a neglect of the deformations of loaded bodies, a direct consequence of the underlying premise that any solid body under consideration is ideally rigid and immutable as to shape and size under all conditions.

Force

A force is any interaction that, when unopposed, will change the motion of an object. A force can cause an object with mass to change its velocity (which includes to begin moving from a state of rest), i.e., to accelerate. Force can also be described intuitively as a push or a pull. A force has both magnitude and direction, making it a vector quantity. It is measured in the SI unit of newtons and represented by the symbol F.

The original form of Newton's second law states that the net force acting upon an object is equal to the rate at which its momentum changes with time. If the mass of the object is constant, this law implies that the acceleration of an object is directly proportional to the net force acting on the object, is in the direction of the net force, and is inversely proportional to the mass of the object.

Concepts related to force include: Thrust, which increases the velocity of an object; drag, which decreases the velocity of an object; and torque, which produces changes in rotational speed of an object. In an extended body, each part usually applies forces on the adjacent parts; the distribution of such forces through the body is the internal mechanical stress. Such internal mechanical stresses cause no acceleration of that body as the forces balance one another. Pressure, the distribution of many small forces applied over an area of a body, is a simple type of stress that if unbalanced can cause the body to accelerate. Stress usually causes deformation of solid materials, or flow in fluids.

Pre-newtonian Concepts

Aristotle famously described a force as anything that causes an object to undergo "unnatural motion".

Since antiquity the concept of force has been recognized as integral to the functioning of each of the simple machines. The mechanical advantage given by a simple machine allowed for less force to be used in exchange for that force acting over a greater distance for the same amount of work. Analysis of the characteristics of forces ultimately culminated in the work of Archimedes who was especially famous for formulating a treatment of buoyant forces inherent in fluids.

Aristotle provided a philosophical discussion of the concept of a force as an integral part of Aristotelian cosmology. In Aristotle's view, the terrestrial sphere contained four elements that come to rest at different "natural places" therein. Aristotle believed that motionless objects on Earth, those composed mostly of the elements earth and water, to be in their natural place on the ground and that they will stay that way if left alone. He distinguished between the innate tendency of objects to find their "natural place" (e.g., for heavy bodies to fall), which led to "natural motion", and unnatural or forced motion, which required continued application of a force. This theory, based on the everyday experience of how objects move, such as the constant application of a force needed to keep a cart moving, had conceptual trouble accounting for the behavior of projectiles, such as the flight of arrows. The place where the archer moves the projectile was at the start of the flight, and while the projectile sailed through the air, no discernible efficient cause acts on it. Aristotle was aware of this problem and proposed that the air displaced through the projectile's path carries the projectile to its target. This explanation demands a continuum like air for change of place in general.

Aristotelian physics began facing criticism in medieval science, first by John Philoponus in the 6th century.

The shortcomings of Aristotelian physics would not be fully corrected until the 17th century work of Galileo Galilei, who was influenced by the late medieval idea that objects in forced motion carried an innate force of impetus. Galileo constructed an experiment in which stones and cannonballs were both rolled down an incline to disprove the Aristotelian theory of motion. He showed that the bodies were accelerated by gravity to an extent that was independent of their mass and argued that objects retain their velocity unless acted on by a force, for example friction.

Newtonian Mechanics

Sir Isaac Newton described the motion of all objects using the concepts of inertia and force, and in doing so he found they obey certain conservation laws. In 1687, Newton published his thesis Philosophiæ Naturalis Principia Mathematica. In this work Newton set out three laws of motion that to this day are the way forces are described in physics.

First Law

Newton's First Law of Motion states that objects continue to move in a state of constant velocity unless acted upon by an external net force (resultant force). This law is an extension of Galileo's insight that constant velocity was associated with a lack of net force. Newton proposed that every object with mass has an innate inertia that functions as the fundamental equilibrium "natural state" in place of the Aristotelian idea of the "natural state of rest". That is, Newton's empirical First Law contradicts the intuitive Aristotelian belief that a net force is required to keep an object moving with constant velocity. By making *rest* physically indistinguishable from *non-zero constant velocity*, Newton's First Law directly connects inertia with the concept of relative velocities. Specifically, in systems where objects are moving with different velocities, it is impossible to determine which object is "in motion" and which object is "at rest". The laws of physics are the same in every inertial frame of reference, that is, in all frames related by a Galilean transformation.

For instance, while traveling in a moving vehicle at a constant velocity, the laws of physics do not change as a result of its motion. If a person riding within the vehicle throws a ball straight up, that person will observe it rise vertically and fall vertically and not have to apply a force in the direction the vehicle is moving. Another person, observing the moving vehicle pass by, would observe the ball follow a curving parabolic path in the same direction as the motion of the vehicle. It is the inertia of the ball associated with its constant velocity in the direction of the vehicle's motion that ensures the ball continues to move forward even as it is thrown up and falls back down. From the perspective of the person in the car, the vehicle and everything inside of it is at rest: It is the outside world that is moving with a constant speed in the opposite direction of the vehicle. Since there is no experiment that can distinguish whether it is the vehicle that is at rest or the outside world that is at rest, the two situations are considered to be physically indistinguishable. Inertia therefore applies equally well to constant velocity motion as it does to rest.

Though Sir Isaac Newton's most famous equation is $\vec{F} = m\vec{a}$, he actually wrote down a different form for his second law of motion that did not use differential calculus.

Second Law

A modern statement of Newton's Second Law is a vector equation:

$$\vec{F} = \frac{d\vec{p}}{dt},$$

where $\vec{p}$ is the momentum of the system, and $\vec{F}$ is the net (vector sum) force. If a body is in equilibrium, there is zero *net* force by definition (balanced forces may be present nevertheless). In contrast, the second law states that if there is an *unbalanced* force acting on an object it will result in the object's momentum changing over time.

By the definition of momentum,

$$\vec{F} = \frac{d\vec{p}}{dt} = \frac{d(m\vec{v})}{dt},$$

where m is the mass and $\vec{v}$ is the velocity.

If Newton's second law is applied to a system of constant mass, m may be moved outside the derivative operator. The equation then becomes:

$$\vec{F} = m\frac{d\vec{v}}{dt}.$$

By substituting the definition of acceleration, the algebraic version of Newton's Second Law is derived:

$$\vec{F} = m\vec{a}.$$

Newton never explicitly stated the formula in the reduced form above.

Newton's Second Law asserts the direct proportionality of acceleration to force and the inverse proportionality of acceleration to mass. Accelerations can be defined through kinematic measurements. However, while kinematics are well-described through reference frame analysis in advanced physics, there are still deep questions that remain as to what is the proper definition of mass. General relativity offers an equivalence between space-time and mass, but lacking a coherent theory of quantum gravity, it is unclear as to how or whether this connection is relevant on microscales. With some justification, Newton's second law can be taken as a quantitative definition of *mass* by writing the law as an equality; the relative units of force and mass then are fixed.

The use of Newton's Second Law as a definition of force has been disparaged in some of the more rigorous textbooks, because it is essentially a mathematical truism. Notable physicists, philosophers and mathematicians who have sought a more explicit definition of the concept of force include Ernst Mach and Walter Noll.

Newton's Second Law can be used to measure the strength of forces. For instance, knowledge of the masses of planets along with the accelerations of their orbits allows scientists to calculate the gravitational forces on planets.

Third Law

Whenever one body exerts a force on another, the latter simultaneously exerts an equal and opposite force on the first. In vector form, if $\vec{F}_{1,2}$ is the force of body 1 on body 2 and $\vec{F}_{2,1}$ that of body 2 on body 1, then:

$$\vec{F}_{1,2} = -\vec{F}_{2,1}.$$

This law is sometimes referred to as the *action-reaction law*, with $\vec{F}_{1,2}$ called the *action* and $-\vec{F}_{2,1}$ the *reaction*.

Newton's Third Law is a result of applying symmetry to situations where forces can be attributed to the presence of different objects. The third law means that all forces are *interactions* between different bodies, and thus that there is no such thing as a unidirectional force or a force that acts on only one body.

In a system composed of object 1 and object 2, the net force on the system due to their mutual interactions is zero:

$$\vec{F}_{1,2} + \vec{F}_{2,1} = 0.$$

More generally, in a closed system of particles, all internal forces are balanced. The particles may accelerate with respect to each other but the center of mass of the system will not accelerate. If an external force acts on the system, it will make the center of mass accelerate in proportion to the magnitude of the external force divided by the mass of the system.

Combining Newton's Second and Third Laws, it is possible to show that the linear momentum of a system is conserved. In a system of two particles, if $\vec{p}_1$ is the momentum of object 1 and $\vec{p}_2$ the momentum of object 2, then:

$$\frac{d\vec{p}_1}{dt} + \frac{d\vec{p}_2}{dt} = \vec{F}_{1,2} + \vec{F}_{2,1} = 0.$$

Using similar arguments, this can be generalized to a system with an arbitrary number of particles. In general, as long as all forces are due to the interaction of objects with mass, it is possible to define a system such that net momentum is never lost nor gained.

Special Theory of Relativity

In the special theory of relativity, mass and energy are equivalent (as can be seen by calculating the work required to accelerate an object). When an object's velocity increases, so does its energy and hence its mass equivalent (inertia). It thus requires more force to accelerate it the same amount than it did at a lower velocity. Newton's Second Law:

$$\vec{F} = \frac{d\vec{p}}{dt}$$

remains valid because it is a mathematical definition. But for relativistic momentum to be conserved, it must be redefined as:

$$\vec{p} = \frac{m_0 \vec{v}}{\sqrt{1 - v^2 / c^2}},$$

where m_0 is the rest mass and c the speed of light.

The relativistic expression relating force and acceleration for a particle with constant non-zero rest mass m moving in the x direction is:

$$\vec{F} = \left(\gamma^3 m a_x, \gamma m a_y, \gamma m a_z\right),$$

Where,

$$\gamma = \frac{1}{\sqrt{1 - v^2 / c^2}}.$$

is called the Lorentz factor.

In the early history of relativity, the expressions $\gamma^3 m$ and γm were called longitudinal and transverse mass. Relativistic force does not produce a constant acceleration, but an ever-decreasing acceleration as the object approaches the speed of light. Note that γ approaches asymptotically an infinite value and is undefined for an object with a non-zero rest mass as it approaches the speed of light, and the theory yields no prediction at that speed.

If v is very small compared to c, then γ is very close to 1 and,

$$F = ma$$

is a close approximation. Even for use in relativity, however, one can restore the form of:

$$F^{\mu} = mA^{\mu}$$

through the use of four-vectors. This relation is correct in relativity when F^{μ} is the four-force, is the invariant mass, and A^{μ} is the four-acceleration.

Descriptions

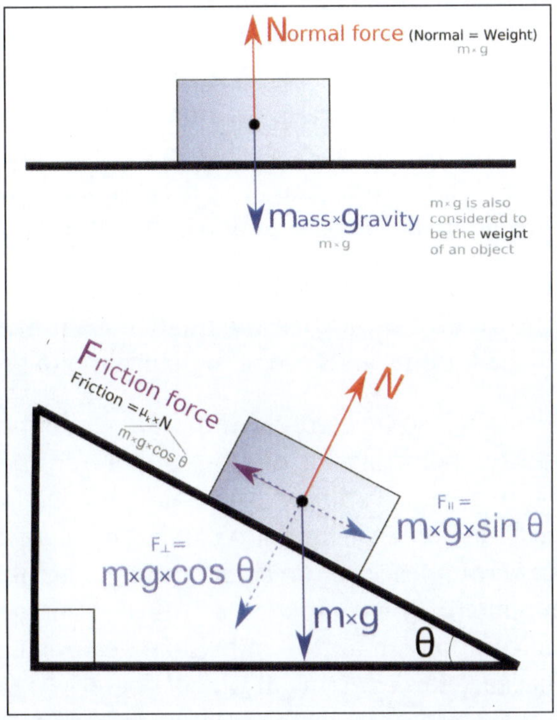

Free body diagrams of a block on a flat surface and an inclined plane. Forces are resolved and added together to determine their magnitudes and the net force.

Since forces are perceived as pushes or pulls, this can provide an intuitive understanding for describing forces. As with other physical concepts (e.g. temperature), the intuitive understanding of forces is quantified using precise operational definitions that are consistent with direct observations and compared to a standard measurement scale. Through experimentation, it is determined that laboratory measurements of forces are fully consistent with the conceptual definition of force offered by Newtonian mechanics.

Forces act in a particular direction and have sizes dependent upon how strong the push or pull is. Because of these characteristics, forces are classified as "vector quantities". This means that forces follow a different set of mathematical rules than physical quantities that do not have direction (denoted scalar quantities). For example, when determining what happens when two forces act on the same object, it is necessary to know both the magnitude and the direction of both forces to calculate the result. If both of these pieces of information are not known for each force, the situation is ambiguous. For example, if you know that two people are pulling on the same rope with known magnitudes of force but you do not know which direction either person is pulling, it is impossible to determine what the acceleration of the rope will be. The two people could be pulling against each other as in tug of war or the two people could be pulling in the same direction. In this simple one-dimensional example, without knowing the direction of the forces it is impossible to decide whether the net force is the result of adding the two force magnitudes or subtracting one from the other. Associating forces with vectors avoids such problems.

Historically, forces were first quantitatively investigated in conditions of static equilibrium where several forces canceled each other out. Such experiments demonstrate the crucial properties that forces are additive vector quantities: they have magnitude and direction. When two forces act on a point particle, the resulting force, the resultant (also called the net force), can be determined by following the parallelogram rule of vector addition: the addition of two vectors represented by sides of a parallelogram, gives an equivalent resultant vector that is equal in magnitude and direction to the transversal of the parallelogram. The magnitude of the resultant varies from the difference of the magnitudes of the two forces to their sum, depending on the angle between their lines of action. However, if the forces are acting on an extended body, their respective lines of application must also be specified in order to account for their effects on the motion of the body.

Free-body diagrams can be used as a convenient way to keep track of forces acting on a system. Ideally, these diagrams are drawn with the angles and relative magnitudes of the force vectors preserved so that graphical vector addition can be done to determine the net force.

As well as being added, forces can also be resolved into independent components at right angles to each other. A horizontal force pointing northeast can therefore be split into two forces, one pointing north, and one pointing east. Summing these component forces using vector addition yields the original force. Resolving force vectors into components of a set of basis vectors is often a more mathematically clean way to describe forces than using magnitudes and directions. This is because, for orthogonal components, the components of the vector sum are uniquely determined by the scalar addition of the components of the individual vectors. Orthogonal components are independent of each other because forces acting at ninety degrees to each other have no effect on the magnitude or direction of the other. Choosing a set of orthogonal basis vectors is often done by considering what set of basis vectors will make the mathematics most convenient. Choosing a basis vector that is in the same direction as one of the forces is desirable, since that force would then have only one non-zero component. Orthogonal force vectors can be three-dimensional with the third component being at right-angles to the other two.

Equilibrium

Equilibrium occurs when the resultant force acting on a point particle is zero (that is, the vector sum of all forces is zero). When dealing with an extended body, it is also necessary that the net torque be zero.

There are two kinds of equilibrium: static equilibrium and dynamic equilibrium.

Static

Static equilibrium was understood well before the invention of classical mechanics. Objects that are at rest have zero net force acting on them.

The simplest case of static equilibrium occurs when two forces are equal in magnitude but opposite in direction. For example, an object on a level surface is pulled (attracted) downward toward the center of the Earth by the force of gravity. At the same time, a force is applied by the surface that resists the downward force with equal upward force (called a normal force). The situation produces zero net force and hence no acceleration.

Pushing against an object that rests on a frictional surface can result in a situation where the object does not move because the applied force is opposed by static friction, generated between the object and the table surface. For a situation with no movement, the static friction force exactly balances the applied force resulting in no acceleration. The static friction increases or decreases in response to the applied force up to an upper limit determined by the characteristics of the contact between the surface and the object.

A static equilibrium between two forces is the most usual way of measuring forces, using simple devices such as weighing scales and spring balances. For example, an object suspended on a vertical spring scale experiences the force of gravity acting on the object balanced by a force applied by the "spring reaction force", which equals the object's weight. Using such tools, some quantitative force laws were discovered: that the force of gravity is proportional to volume for objects of constant density (widely exploited for millennia to define standard weights); Archimedes' principle for buoyancy; Archimedes' analysis of the lever; Boyle's law for gas pressure; and Hooke's law for springs. These were all formulated and experimentally verified before Isaac Newton expounded his Three Laws of Motion.

Dynamic

Galileo Galilei was the first to point out the inherent contradictions contained in Aristotle's description of forces.

Dynamic equilibrium was first described by Galileo who noticed that certain assumptions of Aristotelian physics were contradicted by observations and logic. Galileo realized that simple velocity addition demands that the concept of an "absolute rest frame" did not exist. Galileo concluded that motion in a constant velocity was completely equivalent to rest. This was contrary to Aristotle's notion of a "natural state" of rest that objects with mass naturally approached. Simple experiments showed that Galileo's understanding of the equivalence of constant velocity and rest were correct. For example, if a mariner dropped a cannonball from the crow's nest of a ship moving at a constant velocity, Aristotelian physics would have the cannonball fall straight down while the ship moved beneath it. Thus, in an Aristotelian universe, the falling cannonball would land behind the foot of the mast of a moving ship. However, when this experiment is actually conducted, the cannonball always falls at the foot of the mast, as if the cannonball knows to travel with the ship despite being separated from it. Since there is no forward horizontal force being applied on the cannonball as it falls, the only conclusion left is that the cannonball continues to move with the same velocity as the boat as it falls. Thus, no force is required to keep the cannonball moving at the constant forward velocity.

Moreover, any object traveling at a constant velocity must be subject to zero net force (resultant force). This is the definition of dynamic equilibrium: when all the forces on an object balance but it still moves at a constant velocity.

A simple case of dynamic equilibrium occurs in constant velocity motion across a surface with kinetic friction. In such a situation, a force is applied in the direction of motion while the kinetic friction force exactly opposes the applied force. This results in zero net force, but since the object started with a non-zero velocity, it continues to move with a non-zero velocity. Aristotle misinterpreted this motion as being caused by the applied force. However, when kinetic friction is taken into consideration it is clear that there is no net force causing constant velocity motion.

Forces in Quantum Mechanics

The notion "force" keeps its meaning in quantum mechanics, though one is now dealing with operators instead of classical variables and though the physics is now described by the Schrödinger equation instead of Newtonian equations. This has the consequence that the results of a measurement are now sometimes "quantized", i.e. they appear in discrete portions. This is, of course, difficult to imagine in the context of "forces". However, the potentials V(x,y,z) or fields, from which the forces generally can be derived, are treated similarly to classical position variables, i.e., $V(x,y,z) \rightarrow \hat{V}(\hat{x},\hat{y},\hat{z})$.

This becomes different only in the framework of quantum field theory, where these fields are also quantized.

However, already in quantum mechanics there is one "caveat", namely the particles acting onto each other do not only possess the spatial variable, but also a discrete intrinsic angular momentum-like variable called the "spin", and there is the Pauli exclusion principle relating the space and the spin variables. Depending on the value of the spin, identical particles split into two different classes, fermions and bosons. If two identical fermions (e.g. electrons) have a symmetric spin function (e.g. parallel spins) the spatial variables must be antisymmetric (i.e. they exclude each other from their places much as if there was a repulsive force), and vice versa, i.e. for antiparallel spins the position variables must be symmetric (i.e. the apparent force must be attractive). Thus in the case of two fermions there is a strictly negative correlation between spatial and spin variables, whereas for two bosons (e.g. quanta of electromagnetic waves, photons) the correlation is strictly positive.

Thus the notion "force" loses already part of its meaning.

Feynman Diagrams

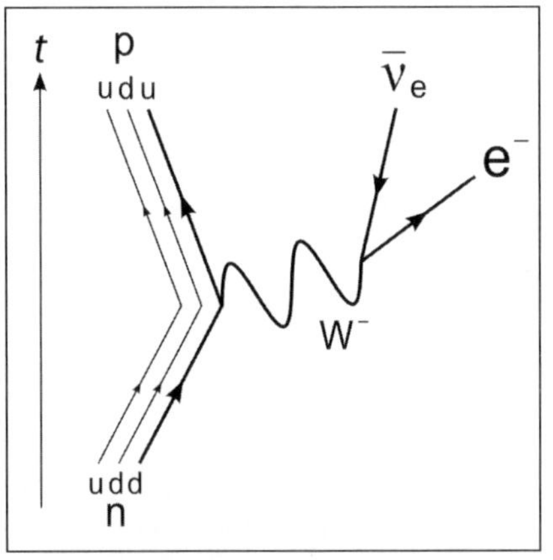

Feynman diagram for the decay of a neutron into a proton. The W boson is between two vertices indicating a repulsion.

In modern particle physics, forces and the acceleration of particles are explained as a mathematical by-product of exchange of momentum-carrying gauge bosons. With the development of quantum field theory and general relativity, it was realized that force is a redundant concept arising from conservation of momentum (4-momentum in relativity and momentum of virtual particles in quantum electrodynamics). The conservation of momentum can be directly derived from the homogeneity or symmetry of space and so is usually considered more fundamental than the concept of a force. Thus the currently known fundamental forces are considered more accurately to be "fundamental interactions". When particle A emits (creates) or absorbs (annihilates) virtual particle B, a momentum conservation results in recoil of particle A making impression of repulsion or attraction between particles A A' exchanging by B. This description applies to all forces arising from fundamental interactions. While sophisticated mathematical descriptions are needed to predict, in full detail, the accurate result of such interactions, there is a conceptually simple way to describe such interactions through the use of Feynman diagrams. In a Feynman diagram, each matter particle is represented as a straight line traveling through time, which normally increases up or to the right in the diagram. Matter and anti-matter particles are identical except for their direction of propagation through the Feynman diagram. World lines of particles intersect at interaction vertices, and the Feynman diagram represents any force arising from an interaction as occurring at the vertex with an associated instantaneous change in the direction of the particle world lines. Gauge bosons are emitted away from the vertex as wavy lines and, in the case of virtual particle exchange, are absorbed at an adjacent vertex.

The utility of Feynman diagrams is that other types of physical phenomena that are part of the general picture of fundamental interactions but are conceptually separate from forces can also be described using the same rules. For example, a Feynman diagram can describe in succinct detail how a neutron decays into an electron, proton, and neutrino, an interaction mediated by the same gauge boson that is responsible for the weak nuclear force.

Fundamental Forces

All of the known forces of the universe are classified into four fundamental interactions. The strong and the weak forces are nuclear forces that act only at very short distances, and are responsible for the interactions between subatomic particles, including nucleons and compound nuclei. The electromagnetic force acts between electric charges, and the gravitational force acts between masses. All other forces in nature derive from these four fundamental interactions. For example, friction is a manifestation of the electromagnetic force acting between atoms of two surfaces, and the Pauli exclusion principle, which does not permit atoms to pass through each other. Similarly, the forces in springs, modeled by Hooke's law, are the result of electromagnetic forces and the Pauli exclusion principle acting together to return an object to its equilibrium position. Centrifugal forces are acceleration forces that arise simply from the acceleration of rotating frames of reference.

The fundamental theories for forces developed from the unification of different ideas. For example, Sir. Isaac Newton unified, with his universal theory of gravitation, the force responsible for objects falling near the surface of the Earth with the force responsible for the falling of celestial bodies about the Earth (the Moon) and around the Sun (the planets). Michael Faraday and James Clerk Maxwell demonstrated that electric and magnetic forces were unified through a theory of electromagnetism. In the 20th century, the development of quantum mechanics led to a modern understanding that the first three fundamental forces (all except gravity) are manifestations of matter (fermions) interacting by exchanging virtual particles called gauge bosons. This standard model of particle physics assumes a similarity between the forces and led scientists to predict the unification of the weak and electromagnetic forces in electroweak theory, which was subsequently confirmed by observation. The complete formulation of the standard model predicts an as yet unobserved Higgs mechanism, but observations such as neutrino oscillations suggest that the standard model is incomplete. A Grand Unified Theory that allows for the combination of the electroweak interaction with the strong force is held out as a possibility with candidate theories such as supersymmetry proposed to accommodate some of the outstanding unsolved problems in physics. Physicists are still attempting to develop self-consistent unification models that would combine all four fundamental interactions into a theory of everything. Einstein tried and failed at this endeavor, but currently the most popular approach to answering this question is string theory.

The four fundamental forces of nature					
Property/Interaction	Gravitation	Weak	Electromagnetic	Strong	
		(Electroweak)		Fundamental	Residual
Acts on:	Mass - Energy	Flavor	Electric charge	Color charge	Atomic nuclei
Particles experiencing:	All	Quarks, leptons	Electrically charged	Quarks, Gluons	Hadrons
Particles mediating:	Graviton (not yet observed)	$W^+ W^- Z^0$	γ	Gluons	Mesons
Strength in the scale of quarks:	10^{-41}	10^{-4}	1	60	Not applicable to quarks
Strength in the scale of protons/neutrons:	10^{-36}	10^{-7}	1	Not applicable to hadrons	20

Gravitational

What we now call gravity was not identified as a universal force until the work of Isaac Newton. Before Newton, the tendency for objects to fall towards the Earth was not understood to be related to the motions of celestial objects. Galileo was instrumental in describing the characteristics of falling objects by determining that the acceleration of every object in free-fall was constant and independent of the mass of the object. Today, this acceleration due to gravity towards the surface of the Earth is usually designated as $\vec{g}$ and has a magnitude of about 9.81 meters per second squared (this measurement is taken from sea level and may vary depending on location), and points toward the center of the Earth. This observation means that the force of gravity on an object at the Earth's surface is directly proportional to the object's mass. Thus an object that has a mass of m will experience a force:

$$\vec{F} = m\vec{g}$$

For an object in free-fall, this force is unopposed and the net force on the object is its weight. For objects not in free-fall, the force of gravity is opposed by the reaction forces applied by their supports. For example, a person standing on the ground experiences zero net force, since a normal force (a reaction force) is exerted by the ground upward on the person that counterbalances his weight that is directed downward.

Newton's contribution to gravitational theory was to unify the motions of heavenly bodies, which Aristotle had assumed were in a natural state of constant motion, with falling motion observed on the Earth. He proposed a law of gravity that could account for the celestial motions using Kepler's laws of planetary motion.

Newton came to realize that the effects of gravity might be observed in different ways at larger distances. In particular, Newton determined that the acceleration of the Moon around the Earth could be ascribed to the same force of gravity if the acceleration due to gravity decreased as an inverse square law. Further, Newton realized that the acceleration of a body due to gravity is proportional to the mass of the other attracting body. Combining these ideas gives a formula that relates the mass ($m_\oplus$) and the radius ($R_\oplus$) of the Earth to the gravitational acceleration:

$$\vec{g} = -\frac{Gm_\oplus}{R_\oplus^{\ 2}}\hat{r}$$

where the vector direction is given by $\hat{r}$, is the unit vector directed outward from the center of the Earth.

In this equation, a dimensional constant G is used to describe the relative strength of gravity. This constant has come to be known as Newton's Universal Gravitation Constant, though its value was unknown in Newton's lifetime. Not until 1798 was Henry Cavendish able to make the first measurement of using a torsion balance; this was widely reported in the press as a measurement of the mass of the Earth since knowing G could allow one to solve for the Earth's mass given the above equation. Newton, however, realized that since all celestial bodies followed the same laws of motion, his law of gravity had to be universal. Succinctly stated, Newton's Law of Gravitation states that the force on a spherical object of mass m_1 due to the gravitational pull of mass m_2 is:

$$\vec{F} = -\frac{Gm_1m_2}{r^2}\hat{r}$$

where r is the distance between the two objects' centers of mass and $\hat{r}$ is the unit vector pointed in the direction away from the center of the first object toward the center of the second object.

This formula was powerful enough to stand as the basis for all subsequent descriptions of motion within the solar system until the 20th century. During that time, sophisticated methods of perturbation analysis were invented to calculate the deviations of orbits due to the influence of multiple bodies on a planet, moon, comet, or asteroid. The formalism was exact enough to allow mathematicians to predict the existence of the planet Neptune before it was observed.

Instruments like GRAVITY provide a powerful probe for gravity force detection.

Mercury's orbit, however, did not match that predicted by Newton's Law of Gravitation. Some astrophysicists predicted the existence of another planet (Vulcan) that would explain the discrepancies; however no such planet could be found. When Albert Einstein formulated his theory of general relativity (GR) he turned his attention to the problem of Mercury's orbit and found that his theory added a correction, which could account for the discrepancy. This was the first time that Newton's Theory of Gravity had been shown to be inexact.

Since then, general relativity has been acknowledged as the theory that best explains gravity. In GR, gravitation is not viewed as a force, but rather, objects moving freely in gravitational fields travel under their own inertia in straight lines through curved space-time – defined as the shortest space-time path between two space-time events. From the perspective of the object, all motion occurs as if there were no gravitation whatsoever. It is only when observing the motion in a global sense that the curvature of space-time can be observed and the force is inferred from the object's curved path. Thus, the straight line path in space-time is seen as a curved line in space, and it is called the ballistic trajectory of the object. For example, a basketball thrown from the ground moves in a parabola, as it is in a uniform gravitational field. Its space-time trajectory is almost a straight line, slightly curved (with the radius of curvature of the order of few light-years). The time derivative of the changing momentum of the object is what we label as "gravitational force".

Electromagnetic

The electrostatic force was first described in 1784 by Coulomb as a force that existed intrinsically between two charges. The properties of the electrostatic force were that it varied as an inverse square law directed in the radial direction, was both attractive and repulsive (there was intrinsic polarity), was independent of the mass of the charged objects, and followed the superposition principle. Coulomb's law unifies all these observations into one succinct statement.

Subsequent mathematicians and physicists found the construct of the electric field to be useful for determining the electrostatic force on an electric charge at any point in space. The electric field was based on using a hypothetical "test charge" anywhere in space and then using Coulomb's Law to determine the electrostatic force. Thus the electric field anywhere in space is defined as:

$$\vec{E} = \frac{\vec{F}}{q}$$

where q is the magnitude of the hypothetical test charge.

Meanwhile, the Lorentz force of magnetism was discovered to exist between two electric currents. It has the same mathematical character as Coulomb's Law with the proviso that like currents attract and unlike currents repel. Similar to the electric field, the magnetic field can be used to determine the magnetic force on an electric current at any point in space. In this case, the magnitude of the magnetic field was determined to be:

$$B = \frac{F}{I\ell}$$

where I is the magnitude of the hypothetical test current and ℓ is the length of hypothetical wire through which the test current flows. The magnetic field exerts a force on all magnets including, for example, those used in compasses. The fact that the Earth's magnetic field is aligned closely with the orientation of the Earth's axis causes compass magnets to become oriented because of the magnetic force pulling on the needle.

Through combining the definition of electric current as the time rate of change of electric charge, a rule of vector multiplication called Lorentz's Law describes the force on a charge moving in a magnetic field. The connection between electricity and magnetism allows for the description of a unified electromagnetic force that acts on a charge. This force can be written as a sum of the electrostatic force (due to the electric field) and the magnetic force (due to the magnetic field). Fully stated, this is the law:

$$\vec{F} = q(\vec{E} + \vec{v} \times \vec{B})$$

where $\vec{F}$ is the electromagnetic force, q is the magnitude of the charge of the particle, $\vec{E}$ is the electric field, $\vec{v}$ is the velocity of the particle that is crossed with the magnetic field ($\vec{B}$).

The origin of electric and magnetic fields would not be fully explained until 1864 when James Clerk Maxwell unified a number of earlier theories into a set of 20 scalar equations, which were later reformulated into 4 vector equations by Oliver Heaviside and Josiah Willard Gibbs. These "Maxwell Equations" fully described the sources of the fields as being stationary and moving charges, and the interactions of the fields themselves. This led Maxwell to discover that electric and magnetic fields could be "self-generating" through a wave that traveled at a speed that he calculated to be the speed of light. This insight united the nascent fields of electromagnetic theory with optics and led directly to a complete description of the electromagnetic spectrum.

However, attempting to reconcile electromagnetic theory with two observations, the photoelectric effect, and the nonexistence of the ultraviolet catastrophe, proved troublesome. Through the work of leading theoretical physicists, a new theory of electromagnetism was developed using quantum mechanics. This final modification to electromagnetic theory ultimately led to quantum electrodynamics (or QED), which fully describes all electromagnetic phenomena as being mediated by wave–particles known as photons. In QED, photons are the fundamental exchange particle, which described all interactions relating to electromagnetism including the electromagnetic force.

Strong Nuclear

There are two "nuclear forces", which today are usually described as interactions that take place in quantum theories of particle physics. The strong nuclear force is the force responsible for the structural integrity of atomic nuclei while the weak nuclear force is responsible for the decay of certain nucleons into leptons and other types of hadrons.

The strong force is today understood to represent the interactions between quarks and gluons as detailed by the theory of quantum chromodynamics (QCD). The strong force is the fundamental force mediated by gluons, acting upon quarks, antiquarks, and the gluons themselves. The (aptly named) strong interaction is the "strongest" of the four fundamental forces.

The strong force only acts *directly* upon elementary particles. However, a residual of the force is observed between hadrons (the best known example being the force that acts between nucleons in atomic nuclei) as the nuclear force. Here the strong force acts indirectly, transmitted as gluons, which form part of the virtual pi and rho mesons, which classically transmit the nuclear force. The failure of many searches for free quarks has shown that the elementary particles affected are not directly observable. This phenomenon is called color confinement.

Weak Nuclear

The weak force is due to the exchange of the heavy W and Z bosons. Its most familiar effect is beta decay (of neutrons in atomic nuclei) and the associated radioactivity. The word "weak" derives from the fact that the field strength is some 10^{13} times less than that of the strong force. Still, it is stronger than gravity over short distances. A consistent electroweak theory has also been developed, which shows that electromagnetic forces and the weak force are indistinguishable at a temperatures in excess of approximately 10^{15} kelvins. Such temperatures have been probed in modern particle accelerators and show the conditions of the universe in the early moments of the Big Bang.

Non-fundamental Forces

Some forces are consequences of the fundamental ones. In such situations, idealized models can be utilized to gain physical insight.

Normal Force

The normal force is due to repulsive forces of interaction between atoms at close contact. When their electron clouds overlap, Pauli repulsion (due to fermionic nature of electrons) follows resulting in the force that acts in a direction normal to the surface interface between two objects. The normal force, for example, is responsible for the structural integrity of tables and floors as well as being the force that responds whenever an external force pushes on a solid object. An example of the normal force in action is the impact force on an object crashing into an immobile surface.

FN represents the normal force exerted on the object.

Friction

Friction is a surface force that opposes relative motion. The frictional force is directly related to the normal force that acts to keep two solid objects separated at the point of contact. There are two broad classifications of frictional forces: static friction and kinetic friction.

The static friction force (F_{sf}) will exactly oppose forces applied to an object parallel to a surface contact up to the limit specified by the coefficient of static friction (μ_{sf}) multiplied by the normal force (F_N). In other words, the magnitude of the static friction force satisfies the inequality:

$$0 \leq F_{\mathrm{sf}} \leq \mu_{\mathrm{sf}} F_{\mathrm{N}}.$$

The kinetic friction force (F_{kf}) is independent of both the forces applied and the movement of the object. Thus, the magnitude of the force equals:

$$F_{\mathrm{kf}} = \mu_{\mathrm{kf}} F_{\mathrm{N}},$$

where μ_{kf} is the coefficient of kinetic friction. For most surface interfaces, the coefficient of kinetic friction is less than the coefficient of static friction.

Tension

Tension forces can be modeled using ideal strings that are massless, frictionless, unbreakable, and unstretchable. They can be combined with ideal pulleys, which allow ideal strings to switch physical direction. Ideal strings transmit tension forces instantaneously in action-reaction pairs so that if two objects are connected by an ideal string, any force directed along the string by the first object is accompanied by a force directed along the string in the opposite direction by the second object. By connecting the same string multiple times to the same object through the use of a set-up that uses movable pulleys, the tension force on a load can be multiplied. For every string that acts on a load, another factor of the tension force in the string acts on the load. However, even though such machines allow for an increase in force, there is a corresponding increase in the length of string that must be displaced in order to move the load. These tandem effects result ultimately in the conservation of mechanical energy since the work done on the load is the same no matter how complicated the machine.

Elastic Force

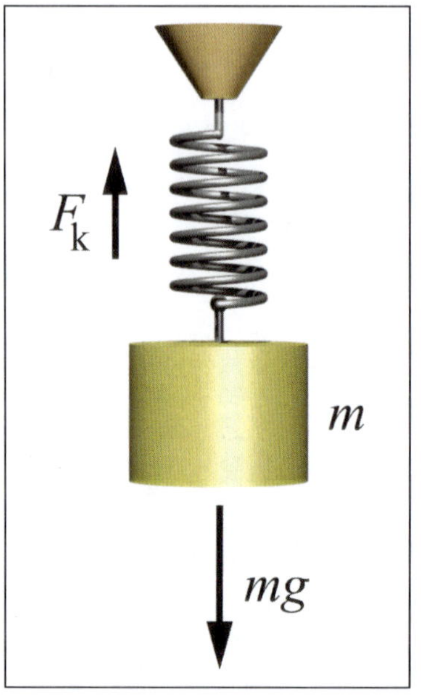

F_k is the force that responds to the load on the spring.

An elastic force acts to return a spring to its natural length. An ideal spring is taken to be massless, frictionless, unbreakable, and infinitely stretchable. Such springs exert forces that push when contracted, or pull when extended, in proportion to the displacement of the spring from its equilibrium position. This linear relationship was described by Robert Hooke in 1676, for whom Hooke's law is named. If Δx is the displacement, the force exerted by an ideal spring equals:

$$\vec{F} = -k\Delta\vec{x}$$

where k is the spring constant (or force constant), which is particular to the spring. The minus sign accounts for the tendency of the force to act in opposition to the applied load.

Continuum Mechanics

Newton's laws and Newtonian mechanics in general were first developed to describe how forces affect idealized point particles rather than three-dimensional objects. However, in real life, matter has extended structure and forces that act on one part of an object might affect other parts of an object. For situations where lattice holding together the atoms in an object is able to flow, contract, expand, or otherwise change shape, the theories of continuum mechanics describe the way forces affect the material.

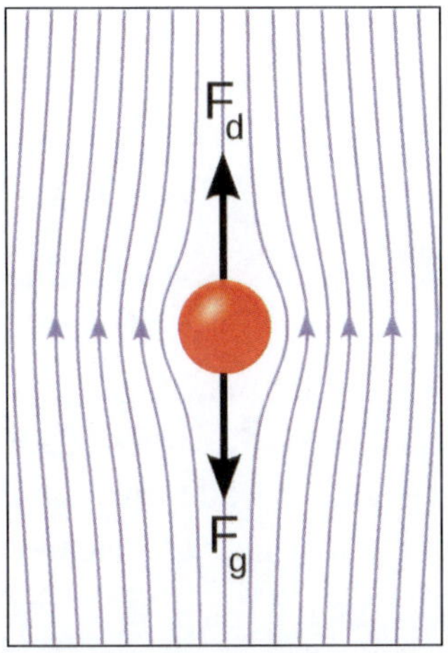

When the drag force (F_d) associated with air resistance becomes equal in magnitude to the force of gravity on a falling object (F_g), the object reaches a state of dynamic equilibrium at terminal velocity.

For example, in extended fluids, differences in pressure result in forces being directed along the pressure gradients as follows:

$$\frac{\vec{F}}{V} = -\vec{\nabla} P$$

where V is the volume of the object in the fluid and P is the scalar function that describes the pressure at all locations in space. Pressure gradients and differentials result in the buoyant force for fluids suspended in gravitational fields, winds in atmospheric science, and the lift associated with aerodynamics and flight.

A specific instance of such a force that is associated with dynamic pressure is fluid resistance: a body force that resists the motion of an object through a fluid due to viscosity. For so-called "Stokes' drag" the force is approximately proportional to the velocity, but opposite in direction:

$$\vec{F}_{\mathrm{d}} = -b\vec{v}$$

where:

is a constant that depends on the properties of the fluid and the dimensions of the object (usually the cross-sectional area),

$\vec{v}$ is the velocity of the object.

More formally, forces in continuum mechanics are fully described by a stress–tensor with terms that are roughly defined as:

$$\sigma = \frac{F}{A}$$

where A is the relevant cross-sectional area for the volume for which the stress-tensor is being calculated. This formalism includes pressure terms associated with forces that act normal to the cross-sectional area (the matrix diagonals of the tensor) as well as shear terms associated with forces that act parallel to the cross-sectional area (the off-diagonal elements). The stress tensor accounts for forces that cause all strains (deformations) including also tensile stresses and compressions.

Fictitious Forces

There are forces that are frame dependent, meaning that they appear due to the adoption of non-Newtonian (that is, non-inertial) reference frames. Such forces include the centrifugal force and the Coriolis force. These forces are considered fictitious because they do not exist in frames of reference that are not accelerating. Because these forces are not genuine they are also referred to as "pseudo forces".

In general relativity, gravity becomes a fictitious force that arises in situations where spacetime deviates from a flat geometry. As an extension, Kaluza–Klein theory and string theory ascribe electromagnetism and the other fundamental forces respectively to the curvature of differently scaled dimensions, which would ultimately imply that all forces are fictitious.

Rotations and Torque

Forces that cause extended objects to rotate are associated with torques. Mathematically, the torque of a force $\vec{F}$ is defined relative to an arbitrary reference point as the cross-product:

$$\vec{\tau} = \vec{r} \times \vec{F}$$

Where,

$\vec{r}$ is the position vector of the force application point relative to the reference point.

Torque is the rotation equivalent of force in the same way that angle is the rotational equivalent for position, angular velocity for velocity, and angular momentum for momentum. As a consequence of Newton's First Law of Motion, there exists rotational inertia that ensures that all bodies maintain their angular momentum unless acted upon by an unbalanced torque. Likewise, Newton's Second Law of Motion can be used to derive an analogous equation for the instantaneous angular acceleration of the rigid body:

$$\vec{\tau} = I\vec{\alpha}$$

Where,

- I is the moment of inertia of the body.
- $\vec{\alpha}$ is the angular acceleration of the body.

This provides a definition for the moment of inertia, which is the rotational equivalent for mass. In more advanced treatments of mechanics, where the rotation over a time interval is described, the moment of inertia must be substituted by the tensor that, when properly analyzed, fully determines the characteristics of rotations including precession and nutation.

Equivalently, the differential form of Newton's Second Law provides an alternative definition of torque:

$\vec{\tau} = \frac{d\vec{L}}{dt}$, where $\vec{L}$ is the angular momentum of the particle.

Newton's Third Law of Motion requires that all objects exerting torques themselves experience equal and opposite torques, and therefore also directly implies the conservation of angular momentum for closed systems that experience rotations and revolutions through the action of internal torques.

Centripetal Force

For an object accelerating in circular motion, the unbalanced force acting on the object equals:

$$\vec{F} = -\frac{mv^2\hat{r}}{r}$$

where m is the mass of the object, v is the velocity of the object and r is the distance to the center of the circular path and $\hat{r}$ is the unit vector pointing in the radial direction outwards from the center. This means that the unbalanced centripetal force felt by any object is always directed toward the center of the curving path. Such forces act perpendicular to the velocity vector associated with the motion of an object, and therefore do not change the speed of the object (magnitude of the velocity), but only the direction of the velocity vector. The unbalanced force that accelerates an object can be resolved into a component that is perpendicular to the path, and one that is tangential to the path. This yields both the tangential force, which accelerates the object by either slowing it down or speeding it up, and the radial (centripetal) force, which changes its direction.

Kinematic Integrals

Forces can be used to define a number of physical concepts by integrating with respect to kinematic variables. For example, integrating with respect to time gives the definition of impulse:

$$\vec{I} = \int_{t_1}^{t_2} \vec{F} dt,$$

Which by Newton's Second Law must be equivalent to the change in momentum (yielding the Impulse momentum theorem).

Similarly, integrating with respect to position gives a definition for the work done by a force:

$$W = \int_{\vec{x}_1}^{\vec{x}_2} \vec{F} \cdot d\vec{x},$$

Which is equivalent to changes in kinetic energy (yielding the work energy theorem).

Power P is the rate of change dW/dt of the work W, as the trajectory is extended by a position change $d\,\vec{x}$ in a time interval dt:

$$dW = \frac{dW}{d\vec{x}} \cdot d\vec{x} = \vec{F} \cdot d\vec{x}, \quad \text{so} \quad P = \frac{dW}{dt} = \frac{dW}{d\vec{x}} \cdot \frac{d\vec{x}}{dt} = \vec{F} \cdot \vec{v},$$

With $\vec{v} = d\vec{x} / dt$ the velocity.

Potential Energy

Instead of a force, often the mathematically related concept of a potential energy field can be used for convenience. For instance, the gravitational force acting upon an object can be seen as the action of the gravitational field that is present at the object's location. Restating mathematically the definition of energy (via the definition of work), a potential scalar field $U(\vec{r})$ is defined as that field whose gradient is equal and opposite to the force produced at every point:

$$\vec{F} = -\vec{\nabla} U.$$

Forces can be classified as conservative or nonconservative. Conservative forces are equivalent to the gradient of a potential while nonconservative forces are not.

Conservative Forces

A conservative force that acts on a closed system has an associated mechanical work that allows energy to convert only between kinetic or potential forms. This means that for a closed system, the net mechanical energy is conserved whenever a conservative force acts on the system. The force, therefore, is related directly to the difference in potential energy between two different locations in space, and can be considered to be an artifact of the potential field in the same way that the direction and amount of a flow of water can be considered to be an artifact of the contour map of the elevation of an area.

Conservative forces include gravity, the electromagnetic force, and the spring force. Each of these forces has models that are dependent on a position often given as a radial vector $\vec{r}$ emanating from spherically symmetric potentials. Examples of this follow:

For gravity:

$$\vec{F}_g = -\frac{Gm_1m_2}{r^2}\hat{r}$$

where, G is the gravitational constant, and m_n is the mass of object n.

For electrostatic forces:

$$\vec{F}_e = \frac{q_1q_2}{4\pi\epsilon_0 r^2}\hat{r}$$

Where ϵ_0 is electric permittivity of free space, and q_n is the electric charge of object n.

For spring forces:

$$\vec{F}_s = -kr\hat{r}$$

where k is the spring constant.

Nonconservative Forces

For certain physical scenarios, it is impossible to model forces as being due to gradient of potentials.

This is often due to macrophysical considerations that yield forces as arising from a macroscopic statistical average of microstates. For example, friction is caused by the gradients of numerous electrostatic potentials between the atoms, but manifests as a force model that is independent of any macroscale position vector. Nonconservative forces other than friction include other contact forces, tension, compression, and drag. However, for any sufficiently detailed description, all these forces are the results of conservative ones since each of these macroscopic forces are the net results of the gradients of microscopic potentials.

The connection between macroscopic nonconservative forces and microscopic conservative forces is described by detailed treatment with statistical mechanics. In macroscopic closed systems, nonconservative forces act to change the internal energies of the system, and are often associated with the transfer of heat. According to the Second law of thermodynamics, nonconservative forces necessarily result in energy transformations within closed systems from ordered to more random conditions as entropy increases.

Units of Measurement

The SI unit of force is the newton (symbol N), which is the force required to accelerate a one kilogram mass at a rate of one meter per second squared, or $kg \cdot m \cdot s^{-2}$. The corresponding CGS unit is the dyne, the force required to accelerate a one gram mass by one centimeter per second squared, or $g \cdot cm \cdot s^{-2}$. A newton is thus equal to 100,000 dynes.

The gravitational foot-pound-second English unit of force is the pound-force (lbf), defined as the force exerted by gravity on a pound-mass in the standard gravitational field of $9.80665\ m \cdot s^{-2}$. The pound-force provides an alternative unit of mass: one slug is the mass that will accelerate by one foot per second squared when acted on by one pound-force.

An alternative unit of force in a different foot-pound-second system, the absolute fps system, is the poundal, defined as the force required to accelerate a one-pound mass at a rate of one foot per second squared. The units of slug and poundal are designed to avoid a constant of proportionality in Newton's Second Law.

The pound-force has a metric counterpart, less commonly used than the newton: the kilogram-force (kgf) (sometimes kilopond), is the force exerted by standard gravity on one kilogram of mass. The kilogram-force leads to an alternate, but rarely used unit of mass: the metric slug (sometimes mug or hyl) is that mass that accelerates at $1\ m \cdot s^{-2}$ when subjected to a force of 1 kgf. The kilogram-force is not a part of the modern SI system, and is generally deprecated; however it still sees use for some purposes as expressing aircraft weight, jet thrust, bicycle spoke tension, torque wrench settings and engine output torque. Other arcane units of force include the sthène, which is equivalent to 1000 N, and the kip, which is equivalent to 1000 lbf.

Units of force					
V·T·E	newton (SI unit)	dyne	kilogram-force, kilopond	pound-force	poundal
1 N	$\equiv 1\ kg \cdot m/s^2$	$= 10^5$ dyn	≈ 0.10197 kp	≈ 0.22481 lbf	≈ 7.2330 pdl
1 dyn	$= 10^{-5}$ N	$\equiv 1\ g \cdot cm/s^2$	$\approx 1.0197 \times 10^{-6}$ kp	$\approx 2.2481 \times 10^{-6}$ lbf	$\approx 7.2330 \times 10^{-5}$ pdl

1 kp	= 9.80665 N	= 980665 dyn	$\equiv g_n \cdot$ (1 kg)	≈ 2.2046 lbf	≈ 70.932 pdl
1 lbf	≈ 4.448222 N	≈ 444822 dyn	≈ 0.45359 kp	$\equiv g_n \cdot$ (1 lb)	≈ 32.174 pdl
1 pdl	≈ 0.138255 N	≈ 13825 dyn	≈ 0.014098 kp	≈ 0.031081 lbf	≡ 1 lb · ft/s²
The value of g_n as used in the official definition of the kilogram-force is used here for all gravitational units.					

Force Systems

A force system is a collection of forces acting at specified locations (may also include couples). Thus the set of forces shown on any free body diagram make up a force system. Force system is simply a term used to describe a group of forces.

Resultants

When two or more forces or moments are combined, the combination is called a resultant. Any force system can be simplified about a point to a resultant consisting of one force and one couple (either or both of which may be zero). The force is applied to the point and is the vector sum of all of the forces in the system. The couple is the vector sum of all of the couples in the force system plus the vector sum of all of the moments about the point of all of the forces in the force system. One immediate consequence of this definition of resultant is that the resultant about any point of the external force system acting on any system in equilibrium must be zero. The statement that the sum of the external forces and the sum of the external moments acting on a system in equilibrium is zero is often replaced by the statement that the resultant acting on a system in equilibrium is zero. An important attribute of the resultant of a force system is that if you apply the resultant of a force system to a rigid body, the effect on that rigid body is exactly the same as the effect of the original force system. It is in this sense that the resultant is "equivalent" to the original force system. Thus as we study the behavior of rigid bodies under the action of forces, the resultant of the external force system will be of vital interest. Care must be taken in replacing force systems with their resultant when dealing with deformable bodies as the effects of the resultant acting on a non-rigid system will be different than the effects of the actual force system.

Equipollence

The "equivalence" of effect on a rigid body of a force system and its resultant is the motivation for the term equipollence. Two force systems are said to be equipollent if they have the same resultant about a point. Thus two equipollent force systems have the same effect on a rigid body. Thus the precise way to state what we have learned about the toe is that the upper clamping surface exerts a force system on the toe that is equipollent to a force of 540 lb in the -Y direction acting at a point at the top, middle of the toe. Typically we are not this precise and merely state that the clamping surface is exerting a force of 540 lb in the -Y direction on the top, middle point of the toe. It is important for you to recognize that when you hear such statements, what they mean is that the force system acting on the object is equipollent to the specified force and that the particular distribution or combination of forces involved may be quite complicated.

Systems of Forces

When a mechanics problem or system has more than one force acting, it is known as a 'force system' or 'system of force'.

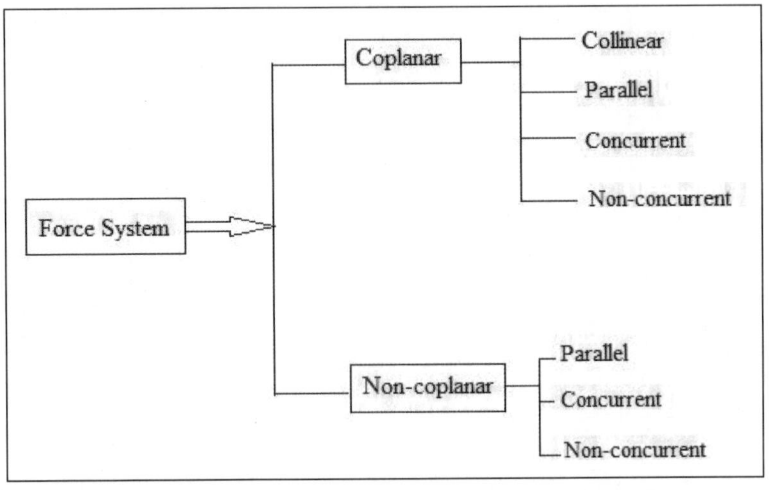

Collinear Force System

When the lines of action of all the forces of a system act along the same line, this force system is called collinear force system.

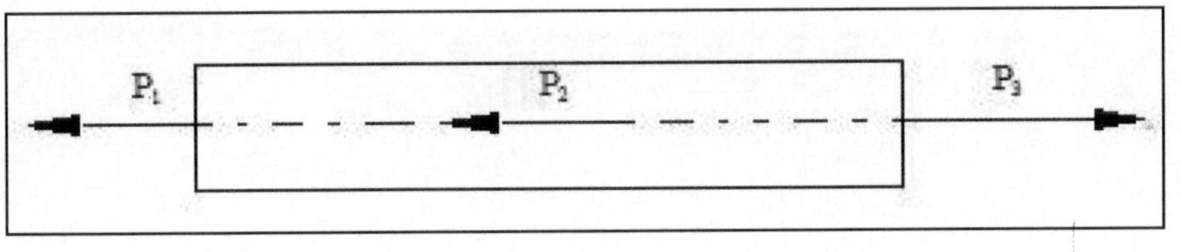

Parallel Forces

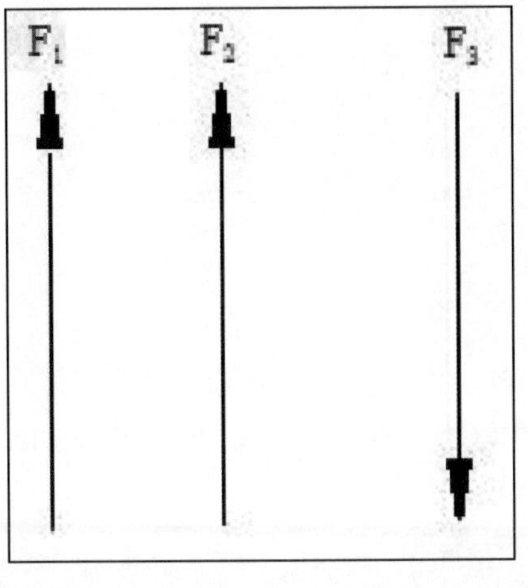

Coplanar Force System

When the lines of action of a set of forces lie in a single plane is called coplanar force system.

Non-coplanar Force System

When the line of action of all the forces do not lie in one plane, is called Non-coplanar force system.

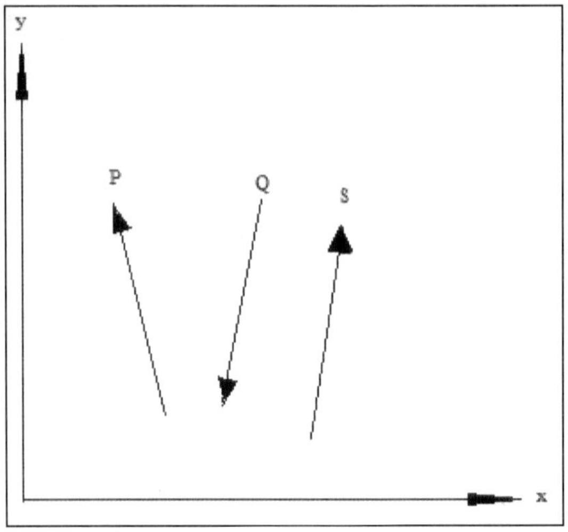

Concurrent Force System

Force is a vector which has both amount and direction. When different forces act on a body and meet at a common point or passes through a same point, then the system is known as concurrent force system.

A body with concurrent forces acting on it is shown in below figure.

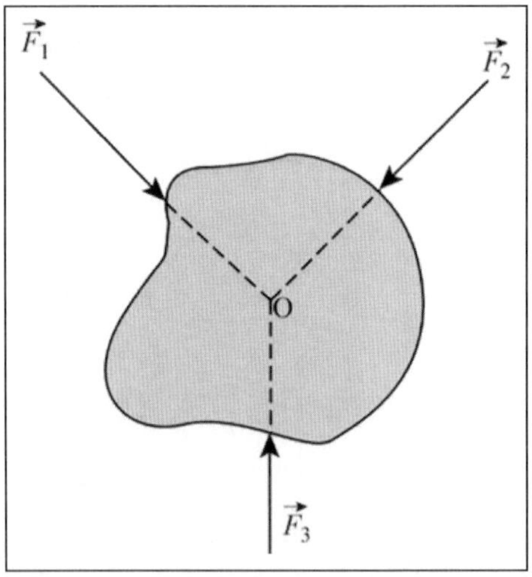

If the forces do not meet at a common point, then there will be an unbalanced moment in the body and the condition of equilibrium will not be satisfied.

In force concurrent systems, equilibrium conditions must be satisfied and it is given below.

$$\sum F_x = 0$$
$$\sum F_y = 0$$

In concurrent force system, moment equilibrium equation is not required because the condition is already satisfied within the system. In the below figure, forces F1 and F2 meets at same point O. It is obvious that the sum of moments produced by these two forces about O will be zero.

Resultant of two concurrent forces is shown in the below figure.

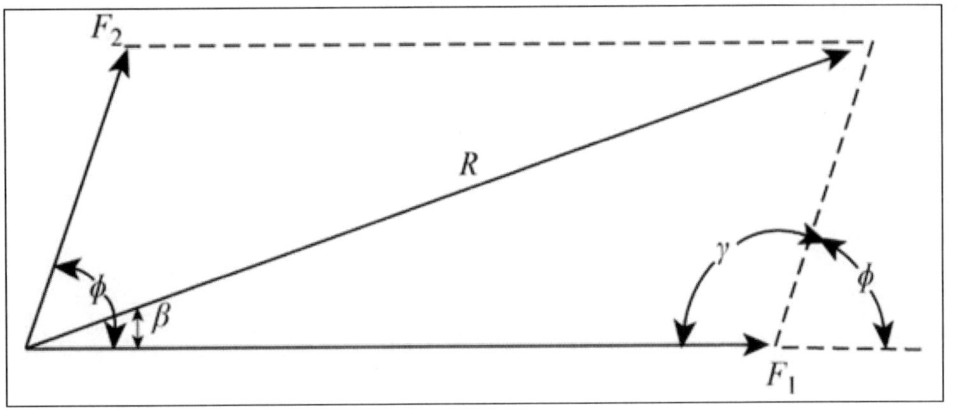

The magnitude R is calculated as below:

$$R = \sqrt{F_1^2 + F_2^2 + 2F_1F_2 \cos\phi}$$

Here, F1 and F2 are concurrent forces and φ is the angle between them.

In a concurrent force system, the below conditions holds good.

- Non parallel forces are concurrent.
- The system is in equilibrium and the equilibrium conditions are satisfied.
- Moment equilibrium equations are not required as the condition of moment equilibrium is already satisfied, due to the concurrent nature of force.

Non-concurrent Force System

When the forces of a system do not meet at a common point of concurrency, this type of force system is called non-concurrent force system. Parallel forces are the example of this type of force system. Non-concurrent forces may be coplanar or non-coplanar.

Coplanar and Concurrent Force System

A force system in which all the forces lie in a single plane and meet at one point, For example, forces acting at a joint of a roof truss.

P = External force,

F_1 to F_5 = Member forces (internal) R_A and R_B = Reactions,

C = Point of concurrency.

Coplanar and Non-concurrent Force System

These forces do not meet at a common point; however, they lie in a single plane, for example, forces acting on a beam.

Non-coplanar and Concurrent Force System

In this system, the forces lie in a different planes but pass through a single point. Example is forces acting at the top end of an electrical pole.

Example: The tension in the guy wires OA and OB of the electrical pole are 500 N and 300 N respectively. Determine the horizontal and vertical components of these tensions exerted by the guy wires on the pole at O.

Solution: The tensions exerted by the guy wires on the pole at O are acting. The components of each of the forces are determined as given in the following table:

Cable	Force P	Inclination with x-axis θ	x-component $Px = P\cos\theta$	y-component $Py = P\sin\theta$
OB	500 N	$\tan^{-1} 6/2 = 71.57°$	500 cos 71.57° = 158.07 N (→)	500 sin 71.57° = 474.36 N (↓)
OA	300 N	$\tan^{-1} 6/1.5 = 75.96°$	300 cos 75.96° = 72.78 N (←)	300 sin 75.96° = 291.04 N (↓)

Non-coplanar and Non-concurrent Force System

The forces which do not lie in a single plane and do not pass through a single point are known as non-coplanar and non-concurrent forces. Example is the loads transferred through columns to the rectangular mat foundation.

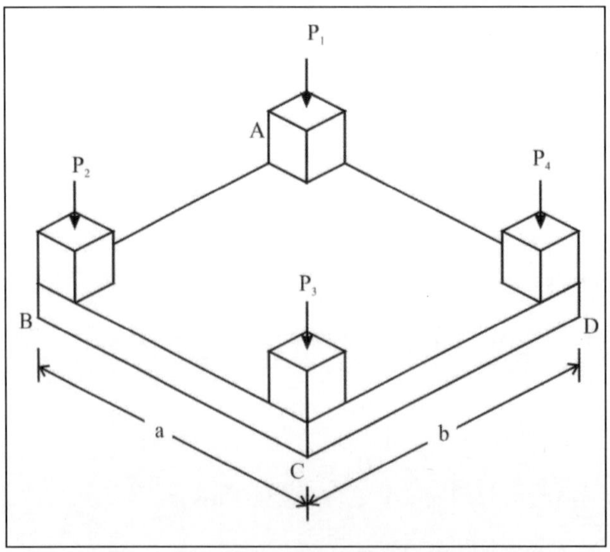

Non-coplanar non-concurrent force system.

Moment of Force

The Moment of a force is a measure of its tendency to cause a body to rotate about a specific point or axis. This is different from the tendency for a body to move, or translate, in the direction of the force. In order for a moment to develop, the force must act upon the body in such a manner that the body would begin to twist. This occurs every time a force is applied so that it does not pass through the centroid of the body. A moment is due to a force not having an equal and opposite force directly along it's line of action.

Imagine two people pushing on a door at the doorknob from opposite sides. If both of them are pushing with an equal force then there is a state of equilibrium. If one of them would suddenly jump back from the door, the push of the other person would no longer have any opposition and the door would swing away. The person who was still pushing on the door created a moment.

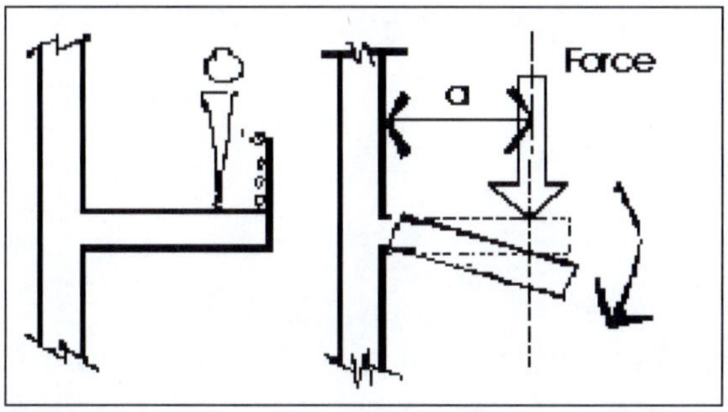

Elements of a Moment

The magnitude of the moment of a force acting about a point or axis is directly proportinoal to the distance of the force from the point or axis. It is defined as the product of the force (F) and the moment arm (d). The moment arm or lever arm is the perpendicular distance between the line of action of the force and the center of moments.

Moment = Force x Distance or M = (F)(d)

The Center of Moments may be the actual point about which the force causes rotation. It may also be a reference point or axis about which the force may be considered as causing rotation. It does not matter as long as a specific point is always taken as the reference point. The latter case is much more common situation in structural design problems.

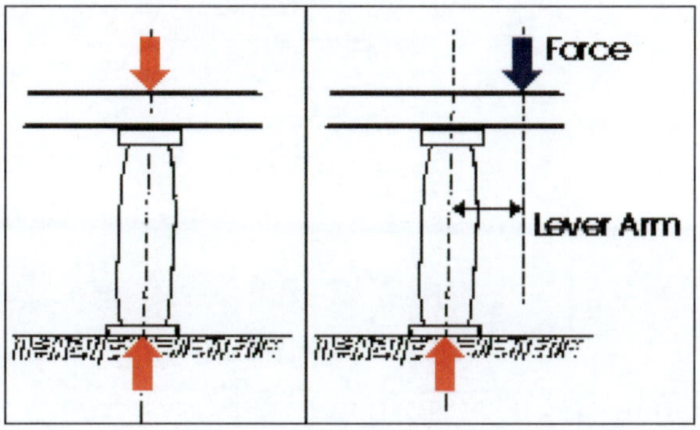

A moment is expressed in units of foot-pounds, kip-feet, newton-meters, or kilonewton-meters. A moment also has a sense; A clockwise rotation about the center of moments will be considered a positive moment; while a counter-clockwise rotation about the center of moments will be considered negative.

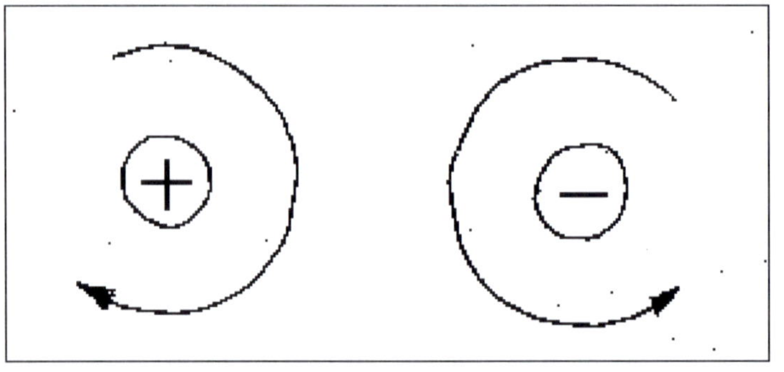

The example shows a wrench being applied to a nut. A 100 pound force is applied to it at point C, the center of the nut. The force is applied at an x- distance of 12 inches from the nut. The center of moments could be point C, but could also be points A or B or D.

Moment about C

The moment arm for calculating the moment around point C is 12 inches. The magnitude of the moment about point C is 12 inches multiplied by the force of 100 lbs to give a total moment of 1200 inch-lbs (or 100 ft-lbs).

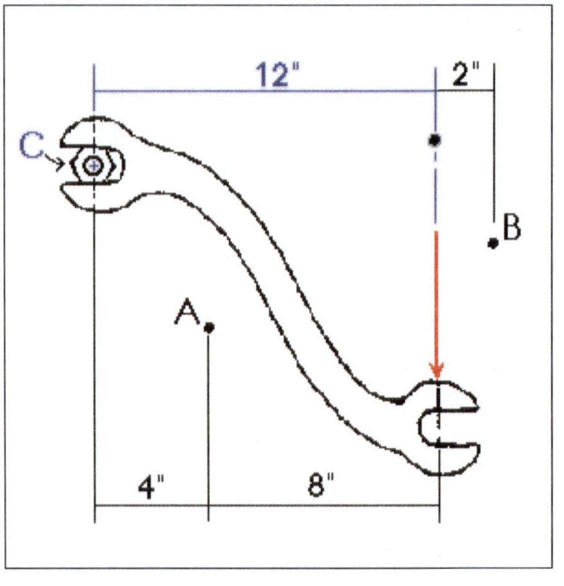

Moment Arm (d) = 12 inches

Magnitude (F) = 100 lbs

Moment = M = 100 lbs x 12 in. = 1200 in-lbs

Similarly, we can find the moments about any point in space.

Moment	A	B	D
Moment Arm	8 inches	2 inches	0 inches
Magnitude of F	100 pounds	100 pounds	100 pounds
Total Moment	800 in- pounds	200 in- pounds	0 in- pounds
	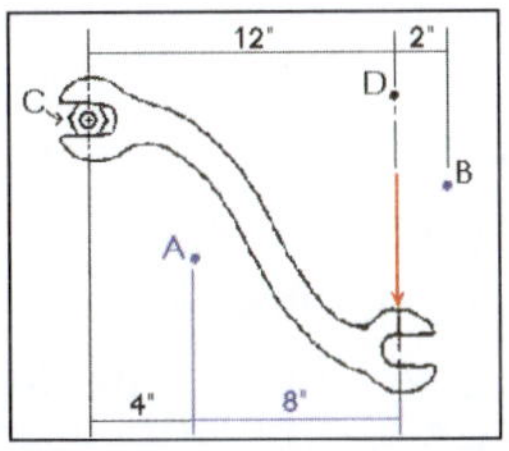	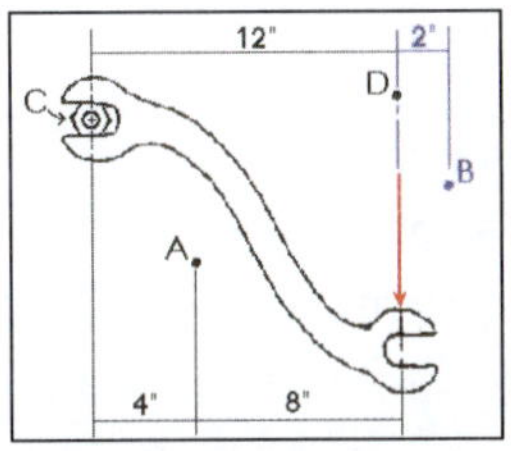	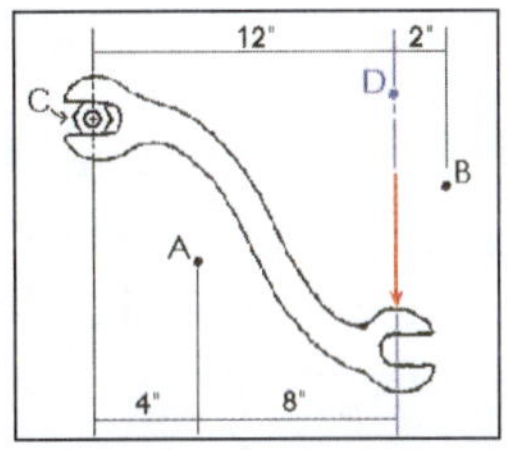

A moment causes a rotation about a point or axis. If the moment is to be taken about a point due to a force F, then in order for a moment to develop, the line of action cannot pass through that point. If the line of action does go through that point, the moment is zero because the magnitude of the moment arm is zero. The total moment was zero because the moment arm was zero as well.

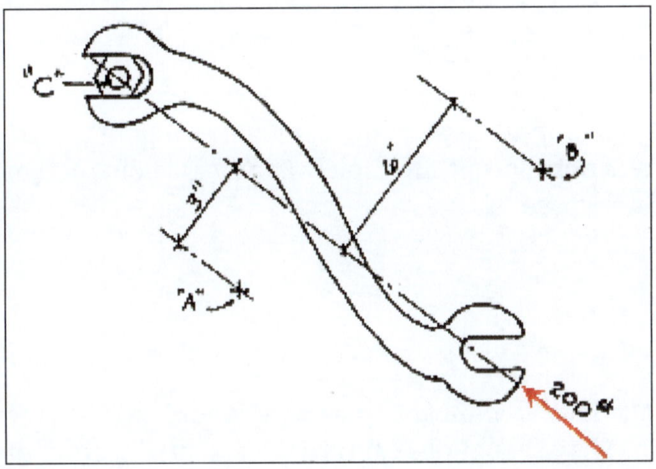

As another example, let us assume that 200 pound force is applied to the wrench as indicated. The moment of the 200 pound force applied at C is zero because:

M = F x d = 200 lbs x 0 in = 0 in-lbs

In other words, there is no tendency for the 200 pound force to cause the wrench to rotate the nut. One could increase the magnitude of the force until the bolt finally broke off (shear failure).

The moment about points X, Y, and Z would also be zero because they also lie on the line of action.

A moment can also be considered to be the result of forces detouring from a direct line drawn between the point of loading of a system and its supports. In this case, the blue force is an eccentric force. In order for it to reach the base of the column, it must make a detour through the beam. The greater the detour, the greater the moment. The most efficient structural systems have the least amount of detours possible.

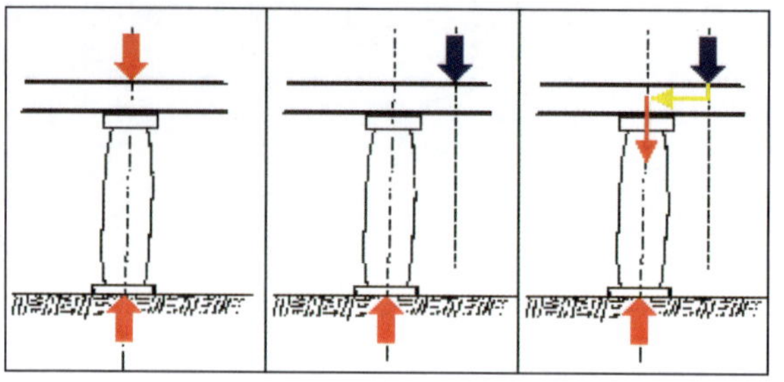

There are cases in which it is easier to calculate the moments of the componenets of a force around a certain point than it is to calculate the moment of the force itself. It could be that the determination of the perpendicular distance of the force is more difficult than determining the perpendicular distance of components of the force. The moment of several forces about a point is simply the algebraic sum of their component moments about the same point. When adding the moments of componenets, one must take great care to be consistant with the sense of each moment.

Elaboration

In its most simple and basic form, a moment is the product of the distance to some point, raised to some power, and some physical quantity such as the force, charge, etc. at that point:

$$\mu_n = r^n Q,$$

Where Q is the physical quantity such as a force applied at a point, or a point charge, or a point mass, etc. If the quantity is not concentrated solely at a single point, the moment is the integral of that quantity's density over space:

$$\mu_n = \int r^n \rho(r)\,dr$$

where ρ is the distribution of the density of charge, mass, or whatever quantity is being considered.

More complex forms take into account the angular relationships between the distance and the physical quantity, but the above equations capture the essential feature of a moment, namely the existence of an underlying $r^n \rho(r)$ or equivalent term. This implies that there are multiple moments (one for each value of n) and that the moment generally depends on the reference point from which the distance r is measured, although for certain moments (technically, the lowest non-zero moment) this dependence vanishes and the moment becomes independent of the reference point.

Each value of n corresponds to a different moment: the 1st moment corresponds to $n = 1$; the 2nd moment to $n = 2$, etc. The 0th moment ($n = 0$) is sometimes called the *monopole moment*; the 1st moment ($n = 1$) is sometimes called the *dipole moment*, and the 2nd moment ($n = 2$) is sometimes called the *quadrupole moment*, especially in the context of electric charge distributions.

Examples:

The moment of force, or torque, is a first moment: $\tau = rF$, or, more generally, $\mathbf{r} \times \mathbf{F}$.

- Similarly, angular momentum is the 1st moment of momentum: $\mathbf{L} = \mathbf{r} \times \mathbf{p}$. Note that momentum itself is not a moment.

The electric dipole moment is also a 1st moment: $\mathbf{p} = q\,\mathbf{d}$ for two opposite point charges or $\int \mathbf{r}\,\rho(\mathbf{r})\,d^3r$ for a distributed charge with charge density $\rho(\mathbf{r})$.

Moments of mass:

- The *total mass* is the zeroth moment of mass.

The center of mass is the 1st moment of mass normalized by total mass: $\mathbf{R} = \frac{1}{M}\sum_i \mathbf{r}_i m_i$ for a collection of point masses, or $\frac{1}{M}\int \mathbf{r}\rho(\mathbf{r})\,d^3r$ for an object with mass distribution $\rho(\mathbf{r})$.

- The moment of inertia is the 2nd moment of mass: $I = r^2 m$ for a point mass, $\sum_i r_i^2 m_i$ for a collection of point masses, or $\int r^2 \rho(\mathbf{r})\,d^3r$ for an object with mass distribution $\rho(\mathbf{r})$. Note that the center of mass is often (but not always) taken as the reference point.

Multipole Moments

Assuming a density function that is finite and localized to a particular region, outside that region a $1/r$ potential may be expressed as a series of spherical harmonics:

$$\Phi(\mathbf{r}) = \int \frac{\rho(\mathbf{r}')}{|\mathbf{r}-\mathbf{r}'|}\,d^3r' = \sum_{\ell=0}^{\infty}\sum_{m=-\ell}^{\ell}\left(\frac{4\pi}{2\ell+1}\right) q_{\ell m}\,\frac{Y_{\ell m}(\theta,\varphi)}{r^{\ell+1}}$$

The coefficients $q_{\ell m}$ are known as *multipole moments*, and take the form:

$$q_{\ell m} = \int (r')^{\ell}\,\rho(\mathbf{r}')\,Y^*_{\ell m}(\theta',\varphi')\,d^3r'$$

where $\mathbf{r}'$ expressed in spherical coordinates (r',φ',θ') is a variable of integration. A more complete treatment may be found in pages describing multipole expansion or spherical multipole moments.

When ρ represents an electric charge density, the q_{lm} are, in a sense, projections of the moments of electric charge: q_{00} is the monopole moment; the q_{1m} are projections of the dipole moment, the q_{2m} are projections of the quadrupole moment, etc.

Applications of Multipole Moments

The multipole expansion applies to $1/r$ scalar potentials, examples of which include the electric potential and the gravitational potential. For these potentials, the expression can be used to approximate the strength of a field produced by a localized distribution of charges (or mass) by calculating the first few moments. For sufficiently large r, a reasonable approximation can be obtained from just the monopole and dipole moments. Higher fidelity can be achieved by calculating higher order moments. Extensions of the technique can be used to calculate interaction energies and intermolecular forces.

The technique can also be used to determine the properties of an unknown distribution ρ. Measurements pertaining to multipole moments may be taken and used to infer properties of the underlying distribution. This technique applies to small objects such as molecules, but has also been applied to the universe itself, being for example the technique employed by the WMAP and Planck experiments to analyze the cosmic microwave background radiation.

Moment of Couples

A couple refers to two parallel forces that are equal in magnitude, opposite in sense and do not share a line of action.

A better term is force couple or pure moment. Its effect is to create rotation without translation, or more generally without any acceleration of the centre of mass. In rigid body mechanics, force couples are free vectors, meaning their effects on a body are independent of torque. This is not to be confused with the term torque as it is used in physics, where it is merely a synonym of moment. Instead, torque is a special case of moment. Torque has special properties that moment does not have, in particular the property of being independent of reference point.

Simple Couple

A couple is a pair of forces, equal in magnitude, oppositely directed, and displaced by perpendicular distance or moment.

The simplest kind of couple consists of two equal and opposite forces whose lines of action do not coincide. This is called a "simple couple". The forces have a turning effect or moment called a torque about an axis which is normal (perpendicular) to the plane of the forces. The SI unit for the torque of the couple is newton metre.

If the two forces are F and –F, then the magnitude of the torque is given by the following formula:

$$\tau = Fd$$

Where,

- τ is the moment of couple,
- F is the magnitude of the force,
- d is the perpendicular distance (moment) between the two parallel forces.

The magnitude of the torque is equal to $F\,d$, with the direction of the torque given by the unit vector $\hat{e}$, which is perpendicular to the plane containing the two forces and positive being a counter-clockwise couple. When d is taken as a vector between the points of action of the forces, then the torque is the cross product of d and F, i.e.

$$\tau = |\mathbf{d} \times \mathbf{F}|.$$

Independence of Reference Point

The moment of a force is only defined with respect to a certain point *P* (it is said to be the "moment about *P*"), and in general when *P* is changed, the moment changes. However, the moment (torque) of a *couple* is *independent* of the reference point *P*: Any point will give the same moment. In other words, a torque vector, unlike any other moment vector, is a "free vector".

The proof of this claim is as follows: Suppose there are a set of force vectors F_1, F_2, etc. that form a couple, with position vectors (about some origin *P*) r_1, r_2, etc., respectively. The moment about *P* is:

$$M = \mathbf{r}_1 \times \mathbf{F}_1 + \mathbf{r}_2 \times \mathbf{F}_2 + \cdots$$

Now we pick a new reference point P' that differs from P by the vector r. The new moment is:

$$M' = (\mathbf{r}_1 + \mathbf{r}) \times \mathbf{F}_1 + (\mathbf{r}_2 + \mathbf{r}) \times \mathbf{F}_2 + \cdots$$

Now the distributive property of the cross product implies:

$$M' = \left(\mathbf{r}_1 \times \mathbf{F}_1 + \mathbf{r}_2 \times \mathbf{F}_2 + \cdots\right) + \mathbf{r} \times \left(\mathbf{F}_1 + \mathbf{F}_2 + \cdots\right).$$

However, the definition of a force couple means that:

$$\mathbf{F}_1 + \mathbf{F}_2 + \cdots = 0.$$

Therefore,

$$M' = \mathbf{r}_1 \times \mathbf{F}_1 + \mathbf{r}_2 \times \mathbf{F}_2 + \cdots = M$$

This proves that the moment is independent of reference point, which is proof that a couple is a free vector.

Forces and Couples

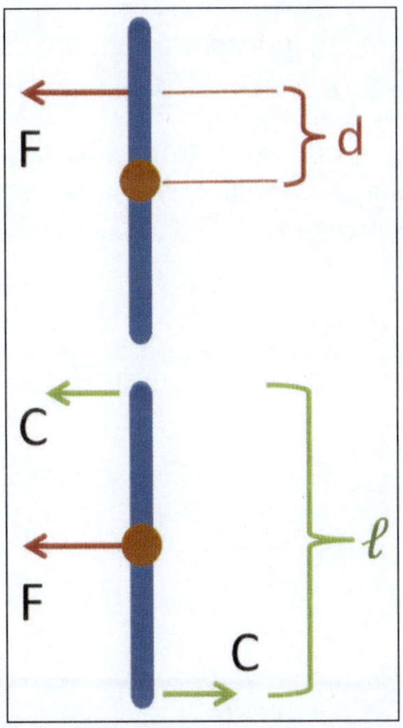

A force *F* applied to a rigid body at a distance *d* from the center of mass has the same effect as the same force applied directly to the center of mass and a couple $C\ell = Fd$. The couple produces an angular acceleration of the rigid body at right angles to the plane of the couple. The force at the center of mass accelerates the body in the direction of the force without change in orientation. The general theorems are:

A single force acting at any point O' of a rigid body can be replaced by an equal and parallel force F acting at any given point O and a couple with forces parallel to F whose moment is $M = Fd$, d being the separation of O and O'. Conversely, a couple and a force in the plane of the couple can be replaced by a single force, appropriately located.

Any couple can be replaced by another in the same plane of the same direction and moment, having any desired force or any desired arm.

Applications

Couples are very important in mechanical engineering and the physical sciences. A few examples are:

- The forces exerted by one's hand on a screw-driver.
- The forces exerted by the tip of a screwdriver on the head of a screw.
- Drag forces acting on a spinning propeller.
- Forces on an electric dipole in a uniform electric field.
- The reaction control system on a spacecraft.
- Force exerted by hands on steering wheel.

In a liquid crystal it is the rotation of an optic axis called the *director* that produces the functionality of these compounds. As Jerald Ericksen explained.

At first glance, it may seem that it is optics or electronics which is involved, rather than mechanics. Actually, the changes in optical behavior, etc. are associated with changes in orientation. In turn, these are produced by couples. Very roughly, it is similar to bending a wire, by applying couples.

Dynamics 4

- Rotational Dynamics
- Linear Dynamical System
- Analytical Dynamics
- Rotordynamics
- Structural Dynamics
- Dynamic Substructuring
- Vehicle Dynamics
- Kinetic Energy
- Kinematics

Dynamics is one of the branches of mechanics that includes the study of forces and their actions on the motion of objects. It includes rotational dynamics, analytical dynamics, rotor dynamics, structural dynamics, vehicle dynamics, etc. All these aspects associated with dynamics have been carefully analyzed in this chapter.

Dynamics is the subdivision of mechanics that is concerned with the motion of material objects in relation to the physical factors that affect them: force, mass, momentum, energy.

Dynamics can be subdivided into kinematics, which describes motion, without regard to its causes, in terms of position, velocity, and acceleration; and kinetics, which is concerned with the effect of forces and torques on the motion of bodies having mass. The foundations of dynamics were laid at the end of the 16th century by Galileo Galilei who, by experimenting with a smooth ball rolling down an inclined plane, derived the law of motion for falling bodies; he was also the first to recognize that force is the cause of changes in the velocity of a body, a fact formulated by Isaac Newton in the 17th century in his second law of motion. This law states that the force acting on a body is equal to the rate of change of the body's momentum.

The Neat Force

Take Newton's first law of motion and break it into two parts. "An object at rest tends to remain at rest and an object in motion tends to continue moving with constant velocity." That long main clause is where statics lives. "Unless acted upon by a net external force." This short subordinate clause is where we find dynamics.

The word net in the phrase *net force* means total, combined, or overall. It's what you get when all things are considered. The word net is related to the word neat. Finding a net value is something like cleaning up a mathematical mess (or at least reducing a mess). This can be written as $\sum \mathbf{F}$ or as F_{net} (using the subscripted word net to make the symbol read more like spoken language and italics to indicate that knowing the magnitude of the force is often all that matters) or other variations like these.

Force is a vector quantity, which means that direction matters. Use positive values for forces that point in the preferred direction and negative values for forces that point in the opposite direction. If a problem is two dimensional, pick two preferred directions at right angles — something like up and to the right. Pick preferred directions that make your life easy. The laws of physics do not care if you call right positive or left positive. Space, in the mathematical sense, is *isotropic*. It measures the same in all directions.

Newton's second law of motion describes how net force, mass, and acceleration are related. Basically, net force causes acceleration and mass resists it. The best way to write that is not with words but with symbols.

$$\sum \mathbf{F} = m\mathbf{a} \text{ or } F_{net} = ma$$

Example:

Take the unexceptional example of an unexceptional bicycle being unexceptionally pedaled down an unexceptional flat, level road in an unexceptional manner. What are the forces acting on the bicycle and rider (together as a whole)?

Start with the obvious. Everything has weight and weight points down. The bicycle is on a solid surface so there's a normal force pointing normal to that surface. The surface is level so the normal direction is up. The rider is pedaling. This means there's some kind of force pushing the bicycle forward. I don't want to overanalyze the situation, so tires resist rolling, the axle may or may not need lubrication, and the air certainly drags on a moving body. Let's make life simple and call all these forces together friction. The rider pushes the bicycle forward and friction pushes backward.

We're ready to make a free body diagram. Draw a box to represent the bicycle and rider. Draw four arrows coming out of the center of the box to represent the four forces acting on the bicycle and rider. Although not always necessary, one should try to draw the arrows with lengths that correspond to the relative magnitudes of the forces. Long arrows for strong forces. Short ones for weak ones.

Start with the easy pair — weight and normal. Nothing is happening in the vertical direction in this scenario. The road is level and the rider isn't performing a stunt. Weight and normal will cancel each other out. Draw one arrow down and another one up and give them the same length.

Finish with the slightly less easy pair — push and friction. Something is happening in the horizontal direction. Motion is happening in the horizontal direction. The bicycle is going somewhere.

Motion isn't what matters. Change in motion is what matters. Is the bicycle accelerating or moving with a constant velocity? Acceleration makes a situation dynamic. Lack of acceleration makes it static. The direction of the net force determines the acceleration. The force pointing in the direction of the net force will be the stronger of the two.

If the bicycle is speeding up, then the rider is pushing the bike forward more than friction is pushing it backward. If the bicycle is moving with a constant velocity, then push and friction are equal. If the bicycle is slowing down, then the force of friction wins over the force pushing the bike forward.

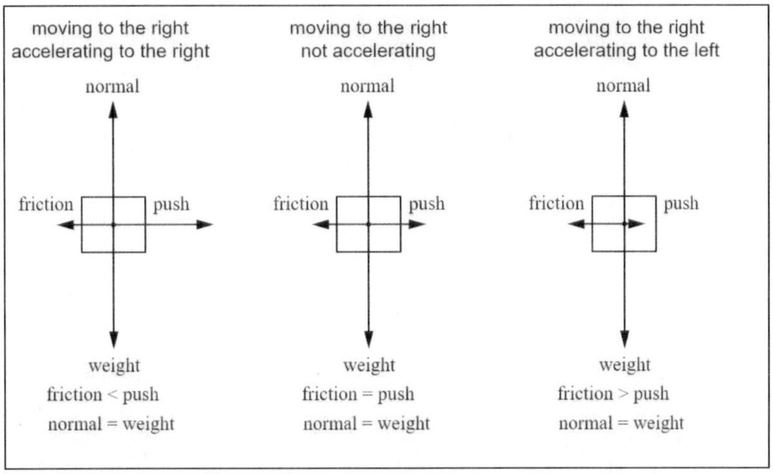

Here's a bunch of equations that show what we just said. We like to use up and right as the positive directions, but that is not a law of physics. It's just a preference.

<u>vertically</u>

$$\Sigma F_y = 0$$

$$N - W = 0$$

$$N = W$$

<u>horizontally</u>

$$\Sigma F_x = ma$$

$$P - f = ma$$

↙ ↓ ↘

<u>speeding up</u>	<u>constant speed</u>	<u>slowing down</u>
$P > f$	$P = f$	$P < f$
$a > 0$	$a = 0$	$a < 0$

Rotational Dynamics

We come across many objects that follow rotational movements. No matter whether fixed or moving, these objects follow special dynamism which lets them perform their specific activity. Whether it is a ceiling fan or a potter's wheel, these rotating objects are a system of particles that consider the motion as a whole. In the introduction to rotational dynamics of a system, we shall emphasize on the center of mass of that particle and use the same in understanding motion as a whole.

Before going deeper into the subject, we should first understand the term "Extended body". When we refer to an object as an extended body we intend to signify it as a system of particles. Rigid bodies are those bodies with definite shape and size. In rigid bodies, the distance between the constituting pairs of particles does not change.

Rotation and its Characteristics

In rotational motion, we know that the particles of the object, while moving follow a circular path. Every particle in the rigid body moves in a circular path along a plane that is perpendicular to the axis and has its center on the same axis. There can be two instances of rotational motion one about the fixed axis and second about an unfixed axis. The examples of rotation around a fixed axis are the fan while for unfixed axis the spinning top makes a perfect example.

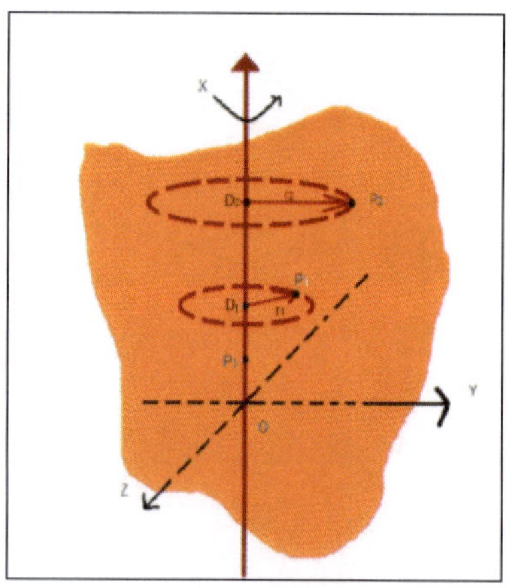

The figure above shows a rigid body's rotation along a fixed axis. Here the axis on which the rotational motion occurs is the X-axis. If P_1 is a particle of the rigid body's system moving with its center D_1 on the fixed axis then the circular plane in which the motion takes place lies in a plane that is perpendicular to the X-axis. The same situation is with another particle P_2. Here both these particles move in a circular motion with their respective radius r_1 and r_2 at centers D_1 and D_2.

The point to be noted here is that though these two particles P1 and P_2 move in a plane perpendicular to the X-axis, their point of rotation is different from each other. Now another particle P_3 seems stationary as its r = 0. Therefore while studying an object's motion, we sum up the movement of various particles to reach a conclusion. In rotational motion, different particles show different points of motion.

The Centre of Mass

Imagine a molecule consisting of j atoms, each of which has a force $\underline{\mathbf{F}}$j acting on it. The subscript j simply identifies the atom under consideration. This force is equal to the atom's mass, m_j, times the acceleration it experiences, d^2r_j/dt^2.

$$\underline{\mathbf{F}}_i = \frac{d^2 m_i \underline{\mathbf{r}}_i}{dt^2}$$

The total force $\underline{\mathbf{F}}$, on the molecule is then the vector sum of these individual forces.

$$\underline{\mathbf{F}} = \sum_1 \underline{\mathbf{F}}_i = \frac{d^2\left(\sum_i m_i \underline{\mathbf{r}}_i\right)}{dt^2}$$

Now $\underline{\mathbf{F}}$ is equal to the total external force on the molecule, since Newton's third law tells us that all the forces exerted by atoms within the molecule on each other must cancel out; equal and opposite reactions. This then only leaves the forces exerted on the atoms by external sources.

Now if we symbolise the molecule's mass by M, we can re-express our equation for the total external force on the molecule by:

$$\underline{\mathbf{F}} = M\frac{d^2\underline{\mathbf{R}}}{dt^2}$$

where $\underline{\mathbf{R}}$ is the position vector of the point through which the external forces effectively act, the centre of mass. The position of the centre of mass is therefore given by:

$$\underline{\mathbf{R}} = \frac{1}{M}\sum_i m_i \underline{r}_i$$

We are now in a position to be more specific about the utility of the centre of mass. First, in the absence of any external forces Newton's first law tells us that this point must be moving with constant velocity $d\underline{\mathbf{R}}/dt$. Now the molecule may well be rotating under these conditions (an external agent must set the rotation in train, but need not be present for the rotation to continue). In such a case it is clear that this rotation must have the centre of mass on the axis of rotation, since otherwise $d\underline{\mathbf{R}}/dt$ will not be constant. Indeed any internal motion in the absence of external forces must have the centre of mass at this "still" point.

The corollary of all this is that if we use the centre of mass as the origin for all our measurements of internal motion, such as rotational motion, we can ignore the motion of the centre of mass itself. Indeed we have already used this trick in our examination of SHM in a diatomic molecule.

Rotational Kinematics

As with linear motion we must develop the kinematic equations for circular motion in order to be able to describe the dynamics of rotation. To do this it is obviously not sensible to use the rectilinear co-ordinates x, y, z.

Since the rotation is about a fixed axis the co-ordinates r and θ will enable us to specify a particle's angular kinematics. Since we will only consider a rigid body only the angle θ will vary with time. This means we need only measure the rate of angular rotation, ω:

$$\omega = \frac{d\theta}{dt}$$

and the rate of change of angular rotation, α:

$$\alpha = \frac{d\omega}{dt} = \frac{d^2\theta}{dt^2}$$

in order to describe the kinematics of such systems. If we measure θ in radians then ω, the analogue of velocity, has units of radians per second and α, the analogue of acceleration, has units of radians per second squared.

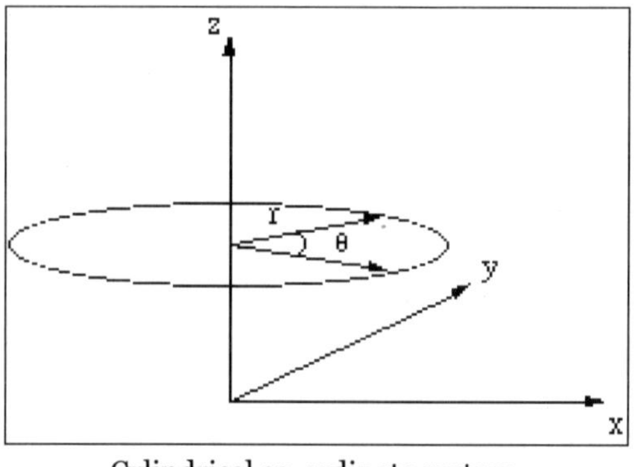

Cylindrical co-ordinate system.

This analogue between one dimensional linear kinematics and angular kinematics, i.e. that θ corresponds to x, ω corresponds to v and α corresponds to a, makes it extremely easy to write down the kinematic equation for motion under constant angular acceleration.

$$\theta = \theta_0 + \omega_0 t + \frac{1}{2}\alpha t^2 \qquad x = x_0 + v_0 t + \frac{1}{2} a t^2$$

$$\omega = \omega_0 + \alpha t \qquad v = v_0 + at$$

$$\omega^2 = \omega_0^2 + 2\alpha(\theta - \theta_0) \qquad v^2 = v_0^2 + 2a(x - x_0)$$

where the subscript zero indicates the value of the appropriate quantity has been measured at t = o. θ, ω and α are actually vectors since they must be able to indicate the direction of the rotational motion. The sensible place to put the vectors **θ**, **ω** and **α** is on the axis of rotation. Having done this the only remaining decision to be made is in which direction the vectors should point. By convention we use the right hand screw rule to determine the direction in which the vectors point.

In the figure the direction of $\underline{\boldsymbol{\theta}}$ and $\underline{\boldsymbol{\omega}}$ have been determined as follows. The particle is rotating anti-clockwise, if we were to turn a right handed screw in the same sense it would travel in the positive z direction. This is the direction in which the vectors must therefore point by the right hand screw rule. We show $\underline{\boldsymbol{\alpha}}$ pointing in the opposite direction, which means that the particle's angular velocity is slowing down.

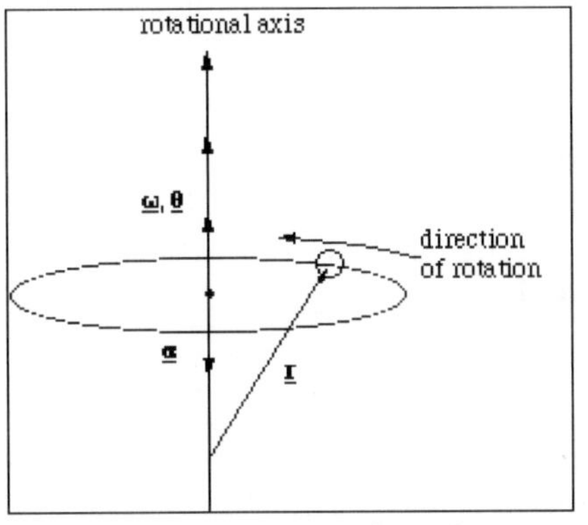

Determining the direction of rotation vectors.

In calculations we will want to be able to relate the angular velocity and acceleration to the instantaneous linear velocity and acceleration of a particle.

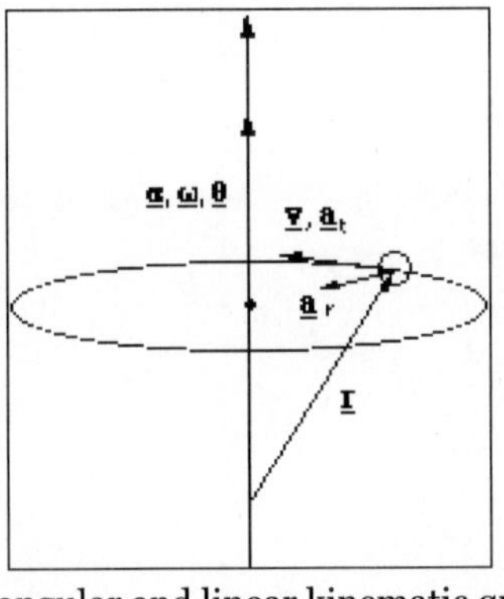

Relating angular and linear kinematic quantities.

The instantaneous velocity is given by:

$$\underline{\mathbf{v}} = \underline{\boldsymbol{\omega}} \times \underline{\mathbf{r}}$$

The order of the vector cross product multiplication, ensures that $\underline{\mathbf{v}}$ points in the correct direction. Remember that the magnitude of $v = \omega r \sin\theta$, where θ is the angle between the vectors $\underline{\boldsymbol{\omega}}$ and $\underline{\mathbf{r}}$. This means that v is simply ω times the perpendicular distance from the axis of rotation to the particle. So even though it make sense to put the origin for the position vector at the centre of mass, for the purposes of these calculations, it suffices to ensure that the origin is somewhere along the axis of rotation.

The radial and tangential accelerations are given by:

$$\underline{\mathbf{a}}_t = \underline{\boldsymbol{\alpha}} \times \underline{\mathbf{r}}$$

$$\underline{\mathbf{a}}_l = \underline{\boldsymbol{\omega}} \times \underline{\mathbf{v}} = \underline{\boldsymbol{\omega}} \times (\underline{\boldsymbol{\omega}} \times \underline{\mathbf{r}})$$

Once again the magnitude of the tangential acceleration is equal to α times the perpendicular distance from the rotation axis to the particle and the radial or centripetal (in towards the centre) acceleration is equal to ω^2 times the perpendicular distance from the axis of rotation to the particle. More importantly, even if the rate of rotation is constant there is still a radial or centripetal acceleration. This must be so, because we are forcing the particle to deviate from a straight path, which according to Newton's first law can only be done by applying a force and hence imparting an acceleration.

Relating Angular and Linear Velocity and Acceleration

In order to examine the dynamics of rotational motion in a molecule we need to determine the instantaneous acceleration and velocity of the constituent atoms. We find the appropriate relations in the following manner.

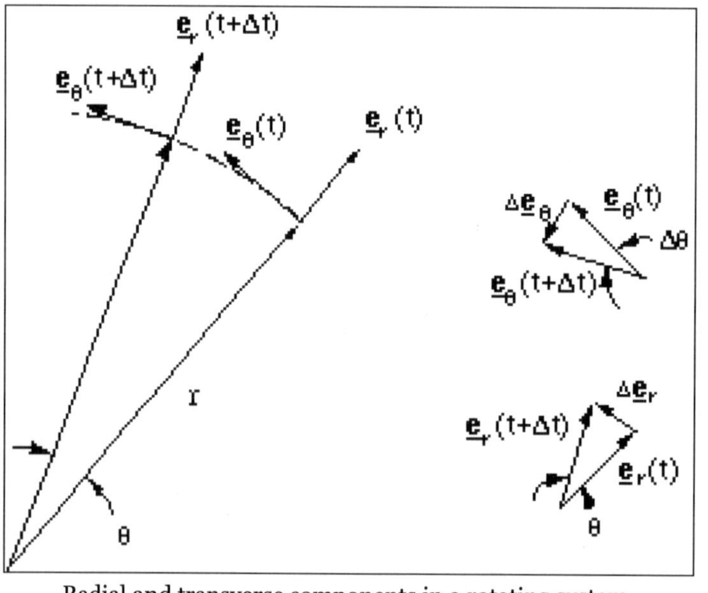

Radial and transverse components in a rotating system.

Figure above shows a particle rotating in a circular path about the z axis in the x-y plane at a constant radius r. We define two units vectors $\underline{e}r$ and $\underline{e}\theta$ pointing in the radial and transverse or tangential directions. So the rectilinear position vector, $\underline{\mathbf{r}}$, is given by:

$$\underline{\mathbf{r}} = \underline{e}r \; r$$

and hence the instantaneous velocity of, $\underline{\mathbf{v}}$, is given by:

$$\frac{d\underline{\mathbf{r}}}{dt} = r\frac{d\underline{\mathbf{e}}_i}{dt}$$

Now looking at our vector diagram we can see that to convert this into an angular velocity ω, we can take advantage of the fact that:

$$\Delta er = \Delta\theta$$

where we define angles in radians, and the angle is small enough, which it will be for an instantaneous measurement. Now examination of the vector diagram should convince you that this change in the radial vector actually points along the transverse direction and hence we get:

$$\frac{d\underline{\mathbf{e}}_i}{dt} = \underline{\mathbf{e}}_\theta \frac{d\theta}{dt} = \underline{\mathbf{e}}_\theta \omega$$

and hence,

$$\underline{\mathbf{v}} = \underline{\mathbf{e}}\theta\ \omega r$$

To find an expression for the acceleration we simply differentiate this expression,

$$\frac{d\underline{\mathbf{v}}}{dt} = \underline{\mathbf{e}}_\theta \frac{d\omega}{dt}\Gamma + \omega = \frac{d\underline{\mathbf{e}}_\theta}{dt} r$$

but dω/dt is simply the angular acceleration α and you can show that,

$$\frac{d\underline{\mathbf{e}}_\theta}{dt} = \underline{\mathbf{e}}_\theta \omega$$

which gives us:

$$\underline{\mathbf{a}} = \underline{\mathbf{e}}\theta\ \alpha r - er\ \omega^2 r$$

This shows us that the acceleration has a tangential and radial component. The tangential acceleration at is simply proportional to the angular acceleration. However, even if α is zero there is a radial acceleration ar, directed towards the axis of rotation (the negative sign). The acceleration is called the centripetal acceleration and is present even at constant angular velocity, because we have to apply a force in order to make a particle deviate from straight line motion (Newton's first law). In the case of circular motion at constant angular velocity, the instantaneous rectilinear velocity is changing in direction, but not magnitude.

We have yet to express the angular kinematic terms as vectors, which we will need to do in order to understand forces and momenta in rotating systems, since these are vectorial. Let us deal with the instantaneous velocity and angular velocity first.

The magnitude of the velocity is given by:

$$v = \omega r$$

Since $\underline{\mathbf{v}}$ and $\underline{\mathbf{r}}$ *are* vectors and ω should also be a vector we can see that the only appropriate combination is the vector cross product, the only remaining ambiguity being the order of the multiplication. By convention the right hand rule is used, giving us:

$$\underline{\mathbf{v}} = \underline{\boldsymbol{\omega} \mathbf{x} \mathbf{r}}$$

Remember that this makes the magnitude of v times the sine of the angle between the vectors w and r. This means that the magnitude of the velocity is equal to the angular velocity times the radial distance between the axis of rotation and the moving particle. The radial and tangential acceleration then follow,

$$\underline{\mathbf{a}}t = \underline{\boldsymbol{\alpha}}\ x\underline{\mathbf{r}}$$

$$\underline{\mathbf{a}}r = \underline{\boldsymbol{\omega}}\ x\ \underline{\mathbf{v}}$$

Torque and Angular Momentum

We now have kinematic equations for rotational motion, so we are equipped to look at the dynamics of rotation. What we will find is that there are direct parallels between the equations of linear dynamics and those for rotational dynamics. This is not a great surprise, since rotational dynamics is in fact only a re-expression of linear dynamics in a rotating frame.

Linear Dynamics Rotational Dynamics

$\underline{\mathbf{v}} / \underline{\boldsymbol{\omega}}$

$\underline{\mathbf{a}} / \underline{\boldsymbol{\alpha}}$

m / I

$\underline{\mathbf{F}} = m\underline{\mathbf{a}} / \underline{\boldsymbol{\tau}} = I\underline{\boldsymbol{\alpha}}$

$\underline{\mathbf{p}} = m\underline{\mathbf{v}} / \underline{\mathbf{L}} = I\underline{\boldsymbol{\omega}}$

dWlin = $\underline{\mathbf{F}} * d\underline{\mathbf{s}}$ / dWrot = $\underline{\boldsymbol{\tau}} * d\underline{\boldsymbol{\theta}}$

Klin = 1/2 mv^2 / Krot = 1/2 $I\omega^2$.

Let us start by looking at the rotational analogue of force, the torque.

Defining the Torque and the Moment of Inertia

Everyday experience tells us that the action which causes an extended object to rotate depends both on the magnitude of the applied force and the distance from the axis of rotation. If you want to push open a door, you apply the force at the edge of the door furthest from the hinge. In this way you maximise the turning effectiveness of your push or pull. If you try opening the door by applying a force close to the hinge you will find the effort required is far greater. Clearer the further away you are from the axis of rotation the greater the effect of a given force. So we would guess that the torque, τ, is related to the force, F, by:

$$\tau = rF$$

where r is the perpendicular distance from the axis of rotation to the point at which the force was applied.

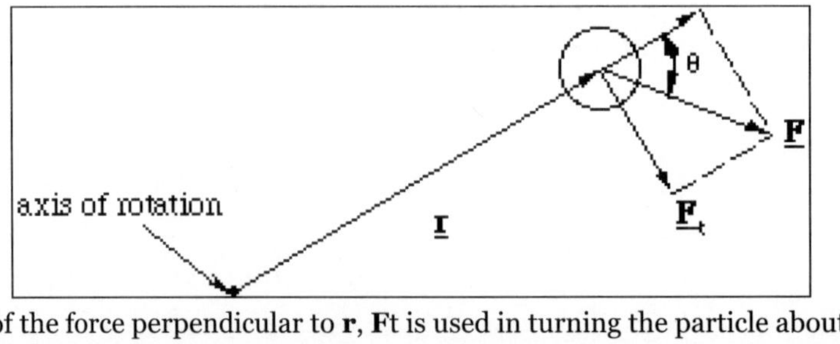

Only the component of the force perpendicular to **r**, **F**t is used in turning the particle about the axis of rotation.

So far we have not determined the vectorial nature of the torque. We can determine this when we note that only the component of force acting tangential, $\underline{\mathbf{F}}$t, to the direction of rotation is used to cause the rotation, figure. Returning to the example of a door, this is simply saying that if you apply the force along the door face you will not make it turn at all, whereas if you apply the force at right angles to the door face you maximise the efficiency with which you turn it.

This means we would now conclude that:

$$\tau = rF \sin\theta$$

where θ is the angle between the position vector and the force vector, figure. Putting this into vectorial language:

$$\underline{\tau} = \underline{r}x\underline{F}$$

and the vectorial construction once again follows the right hand rule. in a molecular calculation the total torque, $\underline{T}$, is the sum of the torque on each atom:

$$\underline{T} = \sum_j \underline{r}_i \times \underline{F}_i$$

where the subscript j identifies the jth atom. Clearly the value of the torque will in general depend on the location we choose as the origin of our co-ordinate system. It is best therefore to choose the centre of mass as the origin for our measurements, although it is well to note that in conditions where the net force on the molecule is zero, i.e. in conditions where the external forces only cause rotational motion, the value of the total torque is equivalent from whatever point the measurements are made.

Since we now have the rotational analogue of the force, we will derive the equation which would be analogous to $\underline{F} = m\underline{a}$. Consider the case of a torque applied to a mass, m, at a distance r from the axis of rotation, where $\theta = \pi/2$.

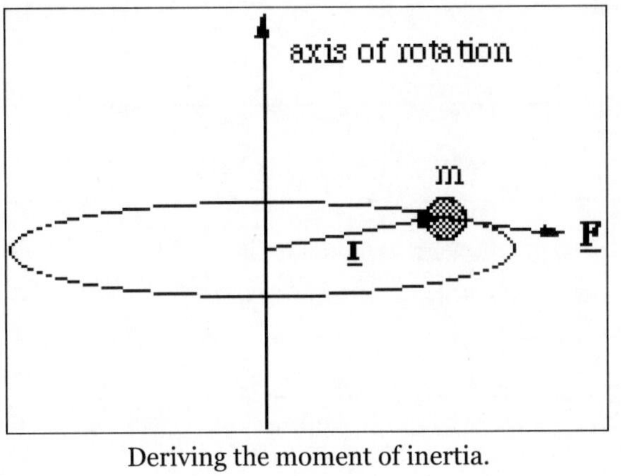

Deriving the moment of inertia.

In this case the magnitude of the torque is:

$$\tau = rF$$

but the applied force, F, is equal to the particle's mass times the tangential acceleration (since r is constant). Therefore,

$$\tau = rma_t$$

but at = αr and therefore,

$$\tau = mr^2\alpha$$

Now mr² is called the moment of inertia, I, which is the rotational analogue of the inertial mass in linear motion. In other words the larger the moment of inertia the harder it is to get the object to change its angular velocity. In full vectorial form we have:

$$\underline{\boldsymbol{\tau}} = \mathrm{I}\,\underline{\boldsymbol{\alpha}}$$

and if you check you can see that this expression is consistent with the direction of vector cross product $\underline{\mathbf{r}} \times \underline{\mathbf{F}}$.

In all the cases of interest to us there will be more than one mass to consider in calculations of the effect of a torque applied to a molecule, so we must generalise the moment of inertia for a system of connected masses (atoms).

$$I = \sum_i m_i r_i^2$$

where rj is the perpendicular distance of the jth atom to the axis of rotation.

Turning a Molecule - The Dipole in an Electrostatic Field

A dipolar molecule in the presence of an electrostatic field will, in general, experience a torque. We will calculate this torque for a diatomic dipolar molecule in a uniform electrostatic field.

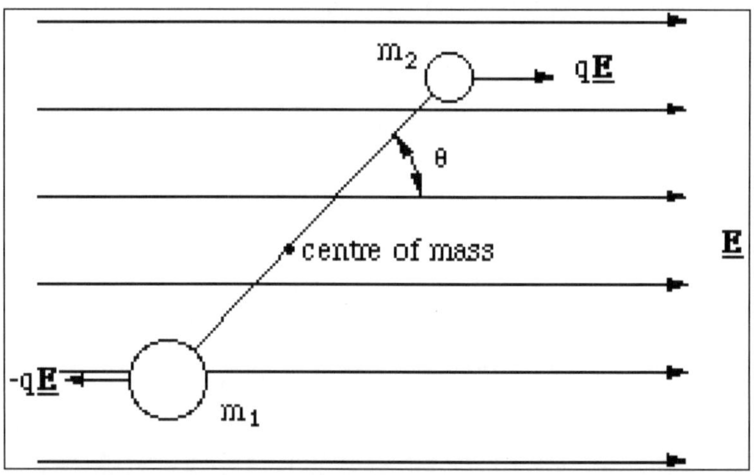

An electrostatic dipole in a uniform electric field. The distance between the dipole charges is 2l.

There is no net force on the dipole, but there is a torque. We will determine the torque exerted by the field on each charge, measured about the centre of mass, using the fact that the magnitude of the torque is given by the product of the distance to the charge and the tangential force. The distance to the centre of mass is given by and hence we find the total torque T to be:

$$\underline{\mathbf{T}} = 2l\frac{m_1}{m_1+m_2}qE\cos(\pi/2-\theta)\hat{\underline{j}} + 2l\frac{m_2}{m_1+m_2}qE\cos(\pi/2-\theta)\hat{\underline{j}}$$

The torque vector is into the paper, since the turning action is clockwise and a right handed screw would travel in this direction. Noting that cos(π/2-θ) = sinθ, reduces to:

$$\underline{\mathbf{T}} = 2lqE\sin\theta\,\hat{j}$$

and we note that $2lq$ is the magnitude of the dipole moment, which suggests that the torque is the vector product of **P** and the field **E**. The appropriate order of vector product is:

$$\underline{\mathbf{T}} = \underline{\mathbf{P}} \times \underline{E}$$

and indeed this is always the torque on an electrostatic dipole in a uniform electric field.

The torque on the dipole depends on sinθ. This means that the force is zero at θ = 0 and π, i.e. the dipole is in equilibrium with P parallel and anti-parallel to E. The difference in these two positions, is that only the parallel state is in stable equilibrium.

Defining the Angular Momentum

We now derive the most important rotational quantity in dynamics, the angular momentum, **L**, the analogue of linear momentum. Now we already know that **F** = d**p**/dt so we should have exactly such a correspondence between the torque and the angular momentum, i.e.

$$\underline{\tau} = \frac{d\underline{\mathbf{L}}}{dt}$$

This expresses the fact that the total angular momentum of a molecule (found by summing the individual momenta of the constituent atoms) is constant unless an external torque is applied. In other words we have identified an important conservation law, that in the absence of any external torques, the angular momentum of an isolated system is always conserved. Put into mathematical terms,

$$\sum_i \underline{\mathbf{L}}_{II} = \sum_i \underline{\mathbf{L}}_{II}$$

where **Lij** is the initial angular momentum of the jth atom and **Lfj** is the final momentum of the jth atom.

Since the torque **τ** = **r** x **F** we anticipate that **L** = **r** x **p**. We can check this by re-expressing **τ** = d**L**/dt as:

$$\underline{\mathbf{r}} \times \underline{\mathbf{F}} = \frac{d}{dt}\left(\underline{\mathbf{r}} \times \underline{\mathbf{p}}\right)$$

and we can confirm the validity of our expression relating torque and angular momentum by doing the differentiation,

$$\frac{d\underline{\mathbf{L}}}{dt} = \frac{d\underline{\mathbf{r}}}{dt} \times m\underline{\mathbf{v}} + \underline{\mathbf{r}} \times m\frac{d\underline{\mathbf{v}}}{dt}$$

$$\frac{d\underline{\mathbf{L}}}{dt} = \underline{\mathbf{v}} + m\underline{\mathbf{v}} \times \underline{\mathbf{r}} \times m\frac{d\underline{\mathbf{v}}}{dt}$$

and noting that $\underline{\mathbf{v}} \times \underline{\mathbf{v}}$ must be zero, which returns us to the original expression and confirms that the angular momentum is given by:

$$\underline{\mathbf{L}} = \underline{\mathbf{r}} \times \underline{\mathbf{p}}$$

Now we can derive the analogue of $\underline{\mathbf{p}} = m\underline{\mathbf{v}}$, which we anticipate will be $\underline{\mathbf{L}} = I\underline{\boldsymbol{\omega}}$, for the angular momentum. For simplicity, but without loss of generality, we set $\underline{\mathbf{r}}$ in the rotational plane of the particle. Recalling that $\underline{\mathbf{v}} = \underline{\boldsymbol{\omega}} \times \underline{\mathbf{r}}$ and substituting into our expression for $\underline{\mathbf{L}}$, we get $\underline{\mathbf{L}} = \underline{\mathbf{r}} \times (m\underline{\boldsymbol{\omega}} \times \underline{\mathbf{r}})$. This has a magnitude $mr^2\omega$ and inspection of the vector diagram, figure, shows that the resultant vector points in the same direction as $\underline{\boldsymbol{\omega}}$, hence:

$$\underline{\mathbf{L}} = I\underline{\boldsymbol{\omega}}$$

So we now see that, as we would have expected, the analogy between linear momentum $\underline{\mathbf{p}} = m\underline{\mathbf{v}}$, and angular momentum is exact (with I and $\underline{\boldsymbol{\omega}}$ replacing m and $\underline{\mathbf{v}}$).

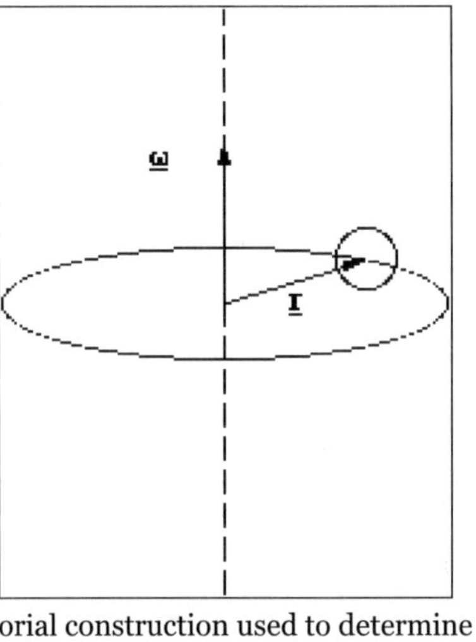

The vectorial construction used to determine l = Iω.

Dipole Oscillations

We have seen how a diatomic molecule stretched along its long axis will oscillate. We now show that a dipolar molecule in the presence of an electric field may also execute rotational oscillations.

In order to perturb the dipole away from the stable equilibrium position at $\theta = 0$ we would have to apply a torque, $\underline{\tau}a$, equal but opposite to the torque due to the field.

$$\underline{\tau} = -PE\sin\theta\,\underline{\hat{j}}$$

According to $\underline{\tau} = \frac{d\underline{\mathbf{L}}}{dt}$ this applied torque is equal to the rate of change in angular momentum, L. Since the bond length does not change, neither can the moment of inertia I, so $d\underline{\mathbf{L}}/dt = I\,d\underline{\boldsymbol{\omega}}/dt = I\,d^2\underline{\boldsymbol{\theta}}/dt^2$ and hence we have,

$$I\frac{d^2\theta}{dt^2} = -PE\sin\theta$$

where I have removed the vector notation since **α** and **θ** point along the same axis. When the perturbation is small sinθ ~ θ and hence we have,

$$\frac{d^2\theta}{dt^2} = -\frac{PE}{I}\theta$$

which is a very simple second order differential equation, the solution of which is $\theta = \theta o \sin(\omega t + \varphi)$, where ω = [[radical]](PE/I) and θo is the angular amplitude. In other words we have simple harmonic oscillations whose period does not depend on the amplitude θo.

Rotational Energetics

Rotational Work, Potential and Kinetic Energy

Finally we will look at the energetics involved in rotating a molecule. Continuing our analogies to linear dynamics, the work done in rotating a molecule through an angle dθ by the action of an external torque will be given by:

$$dWrot = \boldsymbol{\tau} * d\boldsymbol{\theta}$$

which is directly analogues to the definition of work for a linear displacement, with the force being replaced by torque and linear displacement being replaced by angular displacement.

By extension we may then also deduce that the potential energy associated with rotating the molecule is given by:

$$dUrot = -\boldsymbol{\tau} * d\boldsymbol{\theta}$$

The kinetic energy of a particle of mass m moving in a straight line is $1/2\ mv^2$. If the particle moves in a circle we know that v = ωr and hence we would expect the rotational kinetic energy to be:

$$Krot = 1/2\ m\omega^2 r^2 = 1/2\ I\omega^2$$

and once again the moment of inertia replaces the inertial mass and ω replaces v.

We now have the means to solve dynamical problems in rotation using the conservation of energy and the conservation of angular momentum.

Energy of the Oscillating Dipole

Let us return to the case of a dipole in a uniform electric field and determine the energetics of the small rotational oscillations. We first determine the potential energy of the dipole at an angle θ away from stable equilibrium. A torque would be needed to hold the dipole at the angle θ which was equal and opposite to the torque produced by the field, i.e. the holding torque would be -PEsinθ. Hence the potential energy is given by:

$$U(\theta) = \int_0^{\theta} PE \sin\theta d\theta$$

For small perturbations from equilibrium sinθ ~ θ. Using this approximation in the integral gives the following relationship:

$$U(\theta)=\frac{1}{2}PE\theta^2$$

The total energy of the system will include the kinetic energy, $1/2\ I\omega^2$. Hence the total energy H is given by:

$$H=\frac{1}{2}I\left(\frac{d\theta}{dt}\right)^2+\frac{1}{2}PE\theta^2$$

Which has exactly the same mathematical form as the total energy of the simple harmonic oscillator. The dipole will execute simple harmonic oscillations with period T = 2π [[radical]](I/PE).

The Oscillating and Rotating Diatomic Molecule

In general molecules will have translational, vibrational and rotational motion. Since the translational motion can be de coupled from the internal ones, we will now examine a simple case of a homonuclear, diatomic molecule which is both rotating and vibrating. To simplify the problem we have the molecule rotating orthogonally to its molecular axis and vibrating along it.

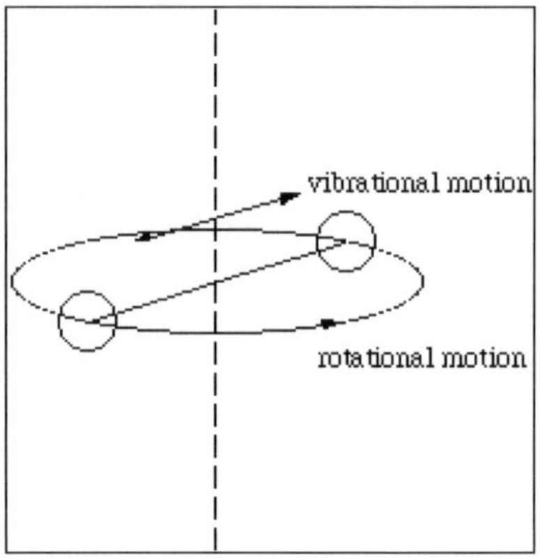

The rotating and vibrating diatomic molecule.

No external forces are applied during the vibrational movement of the atoms from the radial distance r1 to r2. Therefore the initial and final angular momenta, L1 and L2 respectively, must be equal. However, changing the radial separation of the atoms alters the moment of inertia. Hence according to the relationship $L=I\omega$, ω must alter, from ω1 to ω2 to preserve the value of L. For atoms of equal mass, m we then have,

$$2mr_1^2\omega_1=2mr_2^2\omega_2$$

which re-arranges to give:

$$\omega_2=\omega_1\left(\frac{r_1}{r_2}\right)^2$$

where we have omitted the vectorial notation, since the direction of rotation does not alter. Clearly since r1>r2, we have that ω2>ω1. In other words as the atoms come closer together the rotation rate speeds up and as they part company the rotation rate slows down.

It is interesting to note that although the angular momentum is conserved, we can show that the rotational kinetic energy, K, is not.

$$K_2 - K_1 = \frac{1}{2}2mr_2^2\omega_2^2 - \frac{1}{2}2mr_2^2\omega_1^2$$

where K2 is the kinetic energy at position r2 and K1 is that at r1. We simplify this expression by noting that since the angular momentum is conserved we may substitute for ω^2 *by L/2mr²*, to give:

$$K_2 - K_1 = \frac{L^2}{4m}\left(\frac{1}{r_2^2} - \frac{1}{r_1^2}\right) \neq 0$$

which is true for $r1 \neq r2$.

This is actually not too mystifying when you think about it, since moving the atoms inwards takes work. In other words the total energy is conserved, we have just forgotten to take into account the change in potential energy.

The work can be found as follows. An inward radial force, magnitude F, must be exerted to keep the atoms moving in a circular path. This force is equal to the centripetal acceleration times the mass of the atom. For one atom:

$$F = m\omega^2 r$$

Now as the atoms are pulled in the force and displacement are in the same direction and so the work done on one atom is simply:

$$W = \int_{i_1}^{i_2} m\,\omega^2 r\,dr$$

but ω varies with r so we will substitute for ω again using the invariant L= 2mr²ω *(in our case)*. The potential energy change is the negative of the work and remembering that there are two atoms we get:

$$\begin{aligned} U_2 - U_1 &= -2\int_{i_1}^{i_2} m\omega^2 r\,dr \\ &= -\frac{L^2}{2m}\int_{i_1}^{i_2}\frac{1}{r^3}dr \\ &= -\frac{L^2}{4m}\left(\frac{1}{r_1^2} - \frac{1}{r_2^2}\right) \end{aligned}$$

So we now have (U2 - U1) + (K2 - K1) = 0 as required.

The diatomic molecule will clearly change its rate of rotation exactly in time with the natural frequency of the vibrational oscillations. The motions are coupled.

Linear Dynamical System

Linear dynamical systems are dynamical systems whose evaluation functions are linear. While dynamical systems, in general, do not have closed-form solutions, linear dynamical systems can be solved exactly, and they have a rich set of mathematical properties. Linear systems can also be used to understand the qualitative behavior of general dynamical systems, by calculating the equilibrium points of the system and approximating it as a linear system around each such point.

In a linear dynamical system, the variation of a state vector (an N-dimensional vector denoted $\mathbf{x}$) equals a constant matrix (denoted A) multiplied by $\mathbf{x}$. This variation can take two forms: either as a flow, in which $\mathbf{x}$ varies continuously with time:

$$\frac{d}{dt}\mathbf{x}(t) = \mathbf{A} \cdot \mathbf{x}(t)$$

or as a mapping, in which $\mathbf{x}$ varies in discrete steps:

$$\mathbf{x}_{m+1} = \mathbf{A} \cdot \mathbf{x}_m$$

These equations are linear in the following sense: if $\mathbf{x}(t)$ and $\mathbf{y}(t)$ are two valid solutions, then so is any linear combination of the two solutions, e.g., $\mathbf{z}(t) \stackrel{\text{def}}{=} \alpha\mathbf{x}(t) + \beta\mathbf{y}(t)$ where α and β are any two scalars. The matrix $\mathbf{A}$ need not be symmetric.

Linear dynamical systems can be solved exactly, in contrast to most nonlinear ones. Occasionally, a nonlinear system can be solved exactly by a change of variables to a linear system. Moreover, the solutions of (almost) any nonlinear system can be well-approximated by an equivalent linear system near its fixed points. Hence, understanding linear systems and their solutions is a crucial first step to understanding the more complex nonlinear systems.

Solution of Linear Dynamical Systems

If the initial vector $\mathbf{x}_0 \stackrel{\text{def}}{=} \mathbf{x}(t=0)$ is aligned with a right eigenvector $\mathbf{r}_k$ of the matrix $\mathbf{A}$, the dynamics are simple:

$$\frac{d}{dt}\mathbf{x}(t) = \mathbf{A} \cdot \mathbf{r}_k = \lambda_k \mathbf{r}_k$$

where λ_k is the corresponding eigenvalue; the solution of this equation is:

$$\mathbf{x}(t) = \mathbf{r}_k e^{\lambda_k t}$$

as may be confirmed by substitution.

If $\mathbf{A}$ is diagonalizable, then any vector in an N-dimensional space can be represented by a linear combination of the right and left eigenvectors (denoted $\mathbf{l}_k$) of the matrix $\mathbf{A}$.

$$\mathbf{x}_0 = \sum_{k=1}^{N} (\mathbf{l}_k \cdot \mathbf{x}_0)\mathbf{r}_k$$

Therefore, the general solution for $\mathbf{x}(t)$ is a linear combination of the individual solutions for the right eigenvectors:

$$\mathbf{x}(t) = \sum_{k=1}^{n} (\mathbf{l}_k \cdot \mathbf{x}_0)\mathbf{r}_k e^{\lambda_k t}$$

Similar considerations apply to the discrete mappings.

Classification in Two Dimensions

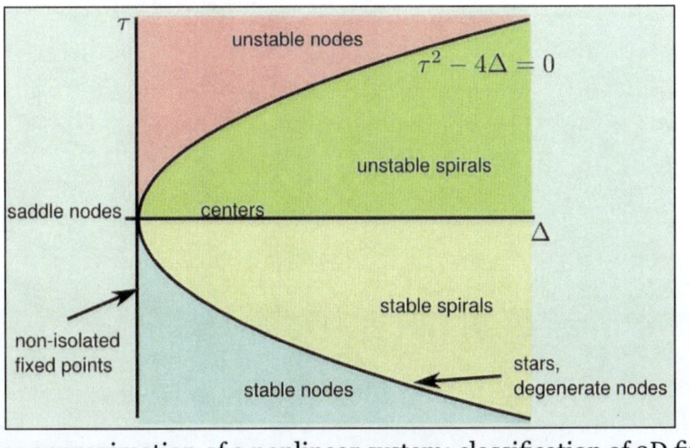

Linear approximation of a nonlinear system: classification of 2D fixed point according to the trace and the determinant of the Jacobian matrix (the linearization of the system near an equilibrium point).

The roots of the characteristic polynomial det(A - λI) are the eigenvalues of A. The sign and relation of these roots, λ_n, to each other may be used to determine the stability of the dynamical system:

$$\frac{d}{dt}\mathbf{x}(t) = \mathbf{A}\mathbf{x}(t).$$

For a 2-dimensional system, the characteristic polynomial is of the form $\lambda^2 - \tau\lambda + \Delta = 0$ where τ is the trace and Δ is the determinant of A. Thus the two roots are in the form:

$$\lambda_1 = \frac{\tau + \sqrt{\tau^2 - 4\Delta}}{2}$$

$$\lambda_2 = \frac{\tau - \sqrt{\tau^2 - 4\Delta}}{2}$$

Note also that $\Delta = \lambda_1\lambda_2$ and $\tau = \lambda_1 + \lambda_2$. Thus if $\Delta < 0$ then the eigenvalues are of opposite sign, and the fixed point is a saddle. If $\Delta > 0$ then the eigenvalues are of the same sign. Therefore, if $\tau > 0$

both are positive and the point is unstable, and if $\tau < 0$ then both are negative and the point is stable. The discriminant will tell you if the point is nodal or spiral (i.e. if the eigenvalues are real or complex).

Analytical Dynamics

Analytical dynamics, or more briefly dynamics, is concerned with the relationship between motion of bodies and its causes, namely the forces acting on the bodies and the properties of the bodies, particularly mass and moment of inertia. The foundation of modern-day dynamics is Newtonian mechanics and its reformulation as Lagrangian mechanics and Hamiltonian mechanics.

The field has a long and important history, as remarked by Hamilton: "The theoretical development of the laws of motion of bodies is a problem of such interest and importance that it has engaged the attention of all the eminent mathematicians since the invention of the dynamics as a mathematical science by Galileo, and especially since the wonderful extension which was given to that science by Newton." Some authors include special relativity within classical dynamics.

Relationship to Statics, Kinetics and Kinematics

Historically, there were three branches of classical mechanics:

- "Statics" (the study of equilibrium and its relation to forces).
- "Kinetics" (the study of motion and its relation to forces).
- "Kinematics" (dealing with the implications of observed motions without regard for circumstances causing them).

These three subjects have been connected to *dynamics* in several ways. One approach combined statics and kinetics under the name dynamics, which became the branch dealing with determination of the motion of bodies resulting from the action of specified forces; another approach separated statics, and combined kinetics and kinematics under the rubric dynamics. This approach is common in engineering books on mechanics, and is still in widespread use among mechanicians.

Fundamental Importance in Engineering, Diminishing Emphasis in Physics

Today, *dynamics* and *kinematics* continue to be considered the two pillars of classical mechanics. Dynamics is still included in mechanical, aerospace, and other engineering curricula because of its importance in machine design, the design of land, sea, air and space vehicles and other applications. However, few modern physicists concern themselves with an independent treatment of "dynamics" or "kinematics," nevermind "statics" or "kinetics." Instead, the entire undifferentiated subject is referred to as *classical mechanics*.

List of Fundamental Dynamics Principles:

- Newton's laws of motion:
  - Inertia.
  - Acceleration.
  - Momentum.
  - Reaction.
- Newton's law of universal gravitation.
- Special theory of relativity.

Rotordynamics

Rotor dynamics is a specialized branch of applied mechanics concerned with the behavior and diagnosis of rotating structures. It is commonly used to analyze the behavior of structures ranging from jet engines and steam turbines to auto engines and computer disk storage. At its most basic level, rotor dynamics is concerned with one or more mechanical structures (rotors) supported by bearings and influenced by internal phenomena that rotate around a single axis. The supporting structure is called a stator. As the speed of rotation increases the amplitude of vibration often passes through a maximum that is called a critical speed. This amplitude is commonly excited by unbalance of the rotating structure; everyday examples include engine balance and tire balance. If the amplitude of vibration at these critical speeds is excessive, then catastrophic failure occurs. In addition to this, turbo machinery often develop instabilities which are related to the internal makeup of turbo machinery, and which must be corrected. This is the chief concern of engineers who design large rotors.

Rotating machinery produces vibrations depending upon the structure of the mechanism involved in the process. Any faults in the machine can increase or excite the vibration signatures. Vibration behavior of the machine due to imbalance is one of the main aspects of rotating machinery. All objects including rotating machinery exhibit natural frequency depending on the structure of the object. The critical speed of a rotating machine occurs when the rotational speed matches its natural frequency. The lowest speed at which the natural frequency is first encountered is called the first critical speed, but as the speed increases, additional critical speeds are seen. Hence, minimizing rotational unbalance and unnecessary external forces are very important to reducing the overall forces which initiate resonance. When the vibration is in resonance, it creates a destructive energy which should be the main concern when designing a rotating machine. The objective here should be to avoid operations that are close to the critical and pass safely through them when in acceleration or deceleration. If this aspect is ignored it might result in loss of the equipment, excessive wear and tear on the machinery, catastrophic breakage beyond repair or even human injury and loss of lives.

The real dynamics of the machine is difficult to model theoretically. The calculations are based on simplified models which resemble various structural components (lumped parameters models), equations obtained from solving models numerically (Rayleigh–Ritz method) and finally from the finite element method (FEM), which is another approach for modelling and analysis of the machine for natural frequencies. There are also some analytical methods, such as the distributed transfer function method, which can generate analytical and closed-form natural frequencies, critical speeds and unbalanced mass response. On any machine prototype it is tested to confirm the precise frequencies of resonance and then redesigned to assure that resonance does not occur.

Basic Principles

The equation of motion, in generalized matrix form, for an axially symmetric rotor rotating at a constant spin speed Ω is:

$$M\ddot{q}(t)+(C+G)\dot{q}(t)+(K+N)q(t) = f(t)$$

where:

- M is the symmetric Mass matrix.
- C is the symmetric damping matrix.
- G is the skew-symmetric gyroscopic matrix.
- K is the symmetric bearing or seal stiffness matrix.
- N is the gyroscopic matrix of deflection for inclusion of e.g., centrifugal elements.

in which q is the generalized coordinates of the rotor in inertial coordinates and f is a forcing function, usually including the unbalance.

The gyroscopic matrix G is proportional to spin speed Ω. The general solution to the above equation involves complex eigenvectors which are spin speed dependent. Engineering specialists in this field rely on the Campbell Diagram to explore these solutions.

An interesting feature of the rotordynamic system of equations are the off-diagonal terms of stiffness, damping, and mass. These terms are called cross-coupled stiffness, cross-coupled damping, and cross-coupled mass. When there is a positive cross-coupled stiffness, a deflection will cause a reaction force opposite the direction of deflection to react the load, and also a reaction force in the direction of positive whirl. If this force is large enough compared with the available direct damping and stiffness, the rotor will be unstable. When a rotor is unstable, it will typically require immediate shutdown of the machine to avoid catastrophic failure.

Campbell Diagram

The Campbell diagram, also known as “Whirl Speed Map” or a “Frequency Interference Diagram”, of a simple rotor system is shown on the right. The pink and blue curves show the backward whirl (BW) and forward whirl (FW) modes, respectively, which diverge as the spin speed increases. When the BW frequency or the FW frequency equal the spin speed Ω, indicated by the intersections

A and B with the synchronous spin speed line, the response of the rotor may show a peak. This is called a critical speed.

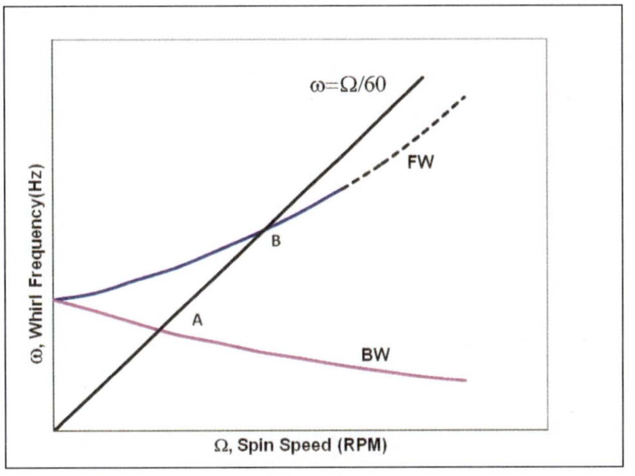

Campbell Diagram for a Simple Rotor.

Jeffcott Rotor

The Jeffcott rotor (named after Henry Homan Jeffcott), also known as the de Laval rotor in Europe, is a simplified lumped parameter model used to solve these equations. The Jeffcott rotor is a mathematical idealization that may not reflect actual rotor mechanics.

Software

There are many software packages that are capable of solving the rotor dynamic system of equations. Rotor dynamic specific codes are more versatile for design purposes. These codes make it easy to add bearing coefficients, side loads, and many other items only a rotordynamicist would need. The non-rotor dynamic specific codes are full featured FEA solvers, and have many years of development in their solving techniques. The non-rotor dynamic specific codes can also be used to calibrate a code designed for rotor dynamics.

Rotordynamic specific codes:

- AxSTREAM RotorDynamics, (SoftInWay) - Integrated Software platform for Rotor Dynamics, capable of lateral, torsional, and axial rotor dynamics for all widely used rotor types using the Finite Element Method on either beam or 2D-axisymmetric elements, and is capable of being automated.
- Dynamics R4 (Alfa-Tranzit Co. Ltd) - Commercial software developed for design and analysis of spatial systems.
- Rotortest, (LAMAR - University of Campinas) - Finite Element Method based software, including different types of bearing solver. Developed by LAMAR (Laboratory of Rotating Machinery) - Unicamp (University of Campinas).
- SAMCEF ROTOR, (SAMCEF) - Software Platform for Rotors Simulation (LMS Samtech,A Siemens Business).

- MADYN (Consulting engineers Klement) - Commercial combined finite element lateral, torsional, axial and coupled solver for multiple rotors and gears, including foundation and housing.
- MADYN 2000 (DELTA JS Inc.) - Commercial combined finite element (3D Timoshenko beam) lateral, torsional, axial and coupled solver for multiple rotors and gears, foundations and casings (capability to import transfer functions and state space matrices from other sources), various bearings (fluid film, spring damper, magnetic, rolling element).
- iSTRDYN (DynaTech Software LLC) - Commercial 2-D Axis-symmetric finite element solver.
- FEMRDYN (DynaTech Engineering, Inc.) - Commercial 1-D Axis-symmetric finite element solver.
- DyRoBeS (Eigen Technologies, Inc.) - Commercial 1-D beam element solver.
- RIMAP (RITEC) - Commercial 1-D beam element solver.
- XLRotor (Rotating Machinery Analysis, Inc.) - Commercial 1-D beam element solver, including magnetic bearing control systems and coupled lateral-torsional analysis. A powerful, fast and easy to use tool for rotor dynamic modeling and analysis using Excel spreadsheets. Readily automated with VBA macros, plus a plugin for 3D CAD software.
- ARMD (Rotor Bearing Technology & Software, Inc.) - Commercial FEA-based software for rotordynamics, multi-branch torsional vibration, fluid-film bearings (hydrodynamic, hydrostatic, and hybrid) design, optimization, and performance evaluation, that is used worldwide by researchers, OEMs and end-users across all industries.
- XLTRC2 (Texas A&M) - Academic 1-D beam element solver.
- ComboRotor (University of Virginia) - Combined finite element lateral, torsional, axial solver for multiple rotors evaluating critical speeds, stability and unbalance response extensively verified by industrial use.
- MESWIR (Institute of Fluid-Flow Machinery, Polish Academy of Sciences) - Academic computer code package for analysis of rotor-bearing systems within the linear and non-linear range.
- RoDAP (D&M Technology) - Commercial lateral, torsional, axial and coupled solver for multiple rotors, gears and flexible disks(HDD).
- ROTORINSA (ROTORINSA) - Commercial finite element software developed by a French engineering school (INSA-Lyon) for analysis of steady-state dynamic behavior of rotors in bending.
- COMSOL Multiphysics, Rotordynamics Module add-on (Rotordynamics Module).
- RAPPID - (Rotordynamics-Seal Research) Commercial finite element based software library (3D solid and beam elements) including rotordynamic coefficient solvers.

Non-rotordynamic specific codes:

- Ansys - Version 11 workbench and classic is capable of solving the rotordynamic equations (3-D/2-D and beam element).
- Nastran - Finite element based (3-D/2-D and beam element).
- SAMCEF - Finite element based (3-D/2-D and beam element).
- eComputingX, Analytical solutions.

Structural Dynamics

Structural dynamics, is a type of structural analysis which covers the behavior of a structure subjected to dynamic (actions having high acceleration) loading. Dynamic loads include people, wind, waves, traffic, earthquakes, and blasts. Any structure can be subjected to dynamic loading. Dynamic analysis can be used to find dynamic displacements, time history, and modal analysis.

Structural analysis is mainly concerned with finding out the behavior of a physical structure when subjected to force. This action can be in the form of load due to the weight of things such as people, furniture, wind, snow, etc. or some other kind of excitation such as an earthquake, shaking of the ground due to a blast nearby, etc. In essence all these loads are dynamic, including the self-weight of the structure because at some point in time these loads were not there. The distinction is made between the dynamic and the static analysis on the basis of whether the applied action has enough acceleration in comparison to the structure's natural frequency. If a load is applied sufficiently slowly, the inertia forces (Newton's first law of motion) can be ignored and the analysis can be simplified as static analysis.

A static load is one which varies very slowly. A dynamic load is one which changes with time fairly quickly in comparison to the structure's natural frequency. If it changes slowly, the structure's response may be determined with static analysis, but if it varies quickly (relative to the structure's ability to respond), the response must be determined with a dynamic analysis.

Dynamic analysis for simple structures can be carried out manually, but for complex structures finite element analysis can be used to calculate the mode shapes and frequencies.

Displacements

A dynamic load can have a significantly larger effect than a static load of the same magnitude due to the structure's inability to respond quickly to the loading (by deflecting). The increase in the effect of a dynamic load is given by the dynamic amplification factor (DAF) or dynamic load factor(DLF):

$$DAF = DLF = \frac{u_{max}}{u_{static}}$$

where u is the deflection of the structure due to the applied load.

Graphs of dynamic amplification factors vs non-dimensional rise time (t_r/T) exist for standard loading functions. Hence the DAF for a given loading can be read from the graph, the static deflection can be easily calculated for simple structures and the dynamic deflection found.

Time History Analysis

A full time history will give the response of a structure over time during and after the application of a load. To find the full time history of a structure's response, you must solve the structure's equation of motion.

Example:

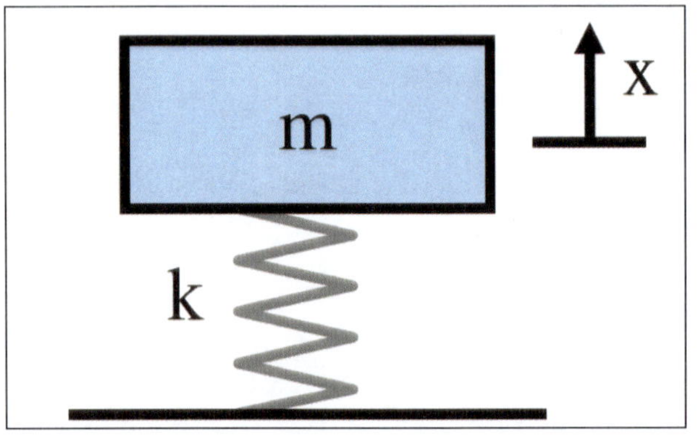

A simple single degree of freedom system (a mass, M, on a spring of stiffness k, for example) has the following equation of motion:

$$M\ddot{x} + kx = F(t)$$

Where $\ddot{x}$ is the acceleration (the double derivative of the displacement) and x is the displacement.

If the loading F(t) is a Heaviside step function (the sudden application of a constant load), the solution to the equation of motion is:

$$x = \frac{F_0}{k}[1 - cos(\omega t)]$$

Where $\omega = \sqrt{\frac{k}{M}}$ and the fundamental natural frequency, $f = \frac{\omega}{2\pi}$.

The static deflection of a single degree of freedom system is:

$$x_{static} = \frac{F_0}{k}$$

so you can write, by combining the above formulae:

$$x = x_{static}[1 - cos(\omega t)]$$

This gives the (theoretical) time history of the structure due to a load F(t), where the false assumption is made that there is no damping.

Although this is too simplistic to apply to a real structure, the Heaviside Step Function is a reasonable model for the application of many real loads, such as the sudden addition of a piece of furniture, or the removal of a prop to a newly cast concrete floor. However, in reality loads are never applied instantaneously - they build up over a period of time (this may be very short indeed). This time is called the rise time.

As the number of degrees of freedom of a structure increases it very quickly becomes too difficult to calculate the time history manually - real structures are analysed using non-linear finite element analysis software.

Damping

Any real structure will dissipate energy (mainly through friction). This can be modelled by modifying the DAF,

$$DAF = 1 + e^{-c\pi}$$

where $c = \dfrac{\text{Damping Coefficient}}{\text{Critical Damping Coefficient}}$ and is typically 2%-10% depending on the type of construction:

- Bolted steel ~6%.
- Reinforced concrete ~ 5%.
- Welded steel ~ 2%.
- Brick masonry ~ 10%.

Methods to Increase Damping

One of the widely used methods to increase damping is to attach a layer of material with a high Damping Coefficient, for example rubber, to a vibrating structure.

Modal Analysis

A modal analysis calculates the frequency modes or natural frequencies of a given system, but not necessarily its full-time history response to a given input. The natural frequency of a system is dependent only on the stiffness of the structure and the mass which participates with the structure (including self-weight). It is not dependent on the load function.

It is useful to know the modal frequencies of a structure as it allows you to ensure that the frequency of any applied periodic loading will not coincide with a modal frequency and hence cause resonance, which leads to large oscillations.

The method is:

- Find the natural modes (the shape adopted by a structure) and natural frequencies.
- Calculate the response of each mode.

- Optionally superpose the response of each mode to find the full modal response to a given loading.

Energy Method

It is possible to calculate the frequency of different mode shape of system manually by the energy method. For a given mode shape of a multiple degree of freedom system you can find an "equivalent" mass, stiffness and applied force for a single degree of freedom system. For simple structures the basic mode shapes can be found by inspection, but it is not a conservative method. Rayleigh's principle states:

"The frequency ω of an arbitrary mode of vibration, calculated by the energy method, is always greater than - or equal to - the fundamental frequency ω_n."

For an assumed mode shape $\bar{u}(x)$, of a structural system with mass M; bending stiffness, EI (Young's modulus, E, multiplied by the second moment of area, I); and applied force, F(x):

$$\text{Equivalent mass, } M_{eq} = \int M\,\bar{u}^2 du$$

$$\text{Equivalent stiffness, } k_{eq} = \int EI\left(\frac{d^2\bar{u}}{dx^2}\right)^2 dx$$

$$\text{Equivalent force, } F_{eq} = \int F\,\bar{u}dx$$

then, as above:

$$\omega = \sqrt{\frac{k_{eq}}{M_{eq}}}$$

Modal Response

The complete modal response to a given load F(x,t) is $v(x,t) = \sum u_n(x,t)$. The summation can be carried out by one of three common methods:

- Superpose complete time histories of each mode (time consuming, but exact).
- Superpose the maximum amplitudes of each mode (quick but conservative).
- Superpose the square root of the sum of squares (good estimate for well-separated frequencies, but unsafe for closely spaced frequencies).

To superpose the individual modal responses manually, having calculated them by the energy method:

Assuming that the rise time t_r is known ($T = 2\pi/\omega$), it is possible to read the DAF from a standard graph. The static displacement can be calculated with $u_{static} = \frac{F_{1,eq}}{k_{1,eq}}$. The dynamic displacement for the chosen mode and applied force can then be found from:

$$u_{max} = u_{static} DAF$$

Modal Participation Factor

For real systems there is often mass participating in the forcing function (such as the mass of ground in an earthquake) and mass participating in inertia effects (the mass of the structure itself, M_{eq}). The modal participation factor Γ is a comparison of these two masses. For a single degree of freedom system $\Gamma = 1$.

$$\Gamma = \frac{\sum M_n \bar{u}_n}{\sum M_n \bar{u}_n^2}$$

Dynamic Substructuring

Dynamic Substructuring (DS) is an engineering tool used to model and analyse the dynamics of mechanical systems by means of its components or substructures. Using the dynamic substructuring approach one is able to analyse the dynamic behaviour of substructures separately and to later on calculate the assembled dynamics using coupling procedures. Dynamic substructuring has several advantages over the analysis of the fully assembled system:

- Substructures can be modelled in the domain that is most appropriate, e.g. experimentally obtained substructures can be combined with numerical models.
- Large and complex systems can be optimized on substructure level.
- Numerical computation load can be reduced as solving several substructures is computationally less demanding than solving one large system.
- Substructure models of different development groups can be shared and combined without exposing the modelling details.

Dynamic substructuring is particularly tailored to simulation of mechanical vibrations, which has implications for many product aspects such as sound / acoustics, fatigue / durability, comfort and safety. Also, dynamic substructuring is applicable to any scale of size and frequency. It is therefore a widely used paradigm in industrial applications ranging from automotive and aerospace engineering to design of wind turbines and high-tech precision machinery.

Domains

Dynamic substructuring can best be seen as a domain-independent toolset for assembly of component models, rather than a modelling method of its own. Generally, dynamic substructuring can be used for all domains that are well suited to simulate multiple input/multiple output behaviour. Five domains that are well suited for substructuring are:

- Physical domain.
- Modal domain.
- Frequency domain.

- Time domain.
- State-space domain.

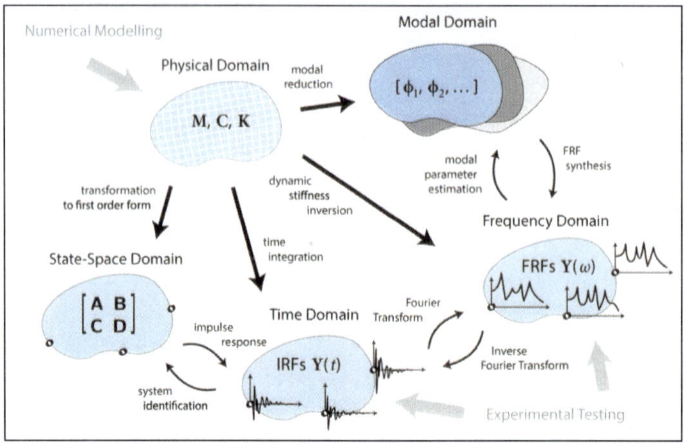

Five domains typically used for dynamic substructuring.

The *physical domain* concerns methods that are based on (linearised) mass, damping and stiffness matrices, typically obtained from numerical FEM modelling. Popular solutions to solve the associated system of second order differential equations are the time integration schemes of Newmark and the Hilbert-Hughes-Taylor scheme. The *modal domain* concerns component mode synthesis (CMS) techniques such as the Craig-Bampton, Rubin and McNeal method. These methods provide efficient modal reduction bases and assembly techniques for numerical models in the physical domain. The *frequency domain* is more popularly known as Frequency Based Substructuring (FBS). Based on the classic formulation of Jetmundsen. and the reformulation of De Klerk, it has become the most commonly used domain for substructuring, because of the ease of expressing the differential equations of a dynamical system (by means of Frequency Response Functions, FRFs) and the convenience of implementing experimentally obtained models. The *time domain* refers to the recently proposed concept of Impulse Based Substructuring (IBS), which expresses the behaviour of a dynamic system using a set of Impulse Response Functions (IRFs). The state-space domain, finally, refers to methods proposed by Sjövall, that employ system identification techniques common to control theory.

Table: Dynamic equations for five domains.

Domain	Dynamic equation	Additional information
Physical domain	$\mathbf{f}(t)=\mathbf{M}\ddot{\mathbf{u}}(t)+\mathbf{C}\dot{\mathbf{u}}(t)+\mathbf{K}\mathbf{u}(t)$	$\mathbf{M},\mathbf{C},\mathbf{K}$ represent the linear(ised) mass, damping and stiffness matrix of the system.
Modal Domain	$\mathbf{f}_{\mathbf{m}}(t)=\mathbf{M}_{\mathbf{m}}\ddot{\eta}(t)+\mathbf{C}_{\mathbf{m}}\dot{\eta}(t)+\mathbf{K}_{\mathbf{m}}\eta(t)$	$\mathbf{M}_m,\mathbf{C}_m,\mathbf{K}_m$ represent the modally reduced mass, damping and stiffness matrix; η is the set of modal amplitudes.
Frequency Domain	$\mathbf{f}(\omega)=\mathbf{Z}(\omega)\mathbf{u}(\omega)$ $\mathbf{u}(\omega)=\mathbf{Y}(\omega)\mathbf{f}(\omega)$	$\mathbf{Z}(\omega)$ is the impedance FRF matrix; $\mathbf{Y}(\omega)$ is the admittance FRF matrix.
Time domain	$\mathbf{u}(t)=\mathbf{Y}(t)*\mathbf{f}(t)$	$\mathbf{Y}(t)$ is the IRF matrix.

State-space domain	$\begin{cases}\dot{\mathbf{x}}(t)=\mathbf{A}\mathbf{x}(t)+\mathbf{B}\mathbf{v}(t)\\ \mathbf{y}(t)=\mathbf{C}\mathbf{x}(t)+\mathbf{D}\mathbf{v}(t)\end{cases}$	$\mathbf{A},\mathbf{B},\mathbf{C},\mathbf{D}$ are the state-space matrices; $\mathbf{x}$, $\mathbf{v}$ and $\mathbf{y}$ represent the state, input and output vector.

As dynamic substructuring is a domain-independent toolset, it is applicable to the dynamic equations of all domains. In order to establish substructure assembly in a particular domain, two interface conditions need to be implemented. This is explained next, followed by a few common substructuring techniques.

Interface Conditions

To establish substructuring coupling/decoupling in each of the domains, two conditions should be met:

- Coordinate compatibility, i.e. the connecting nodes of two substructures should have equal interface displacement.
- Force equilibrium, i.e. the interface forces between connecting nodes have equal magnitude and opposing sign.

These are the two essential conditions that keep substructures together, hence allow to construct an assembly of multiple components. Note that the conditions are comparable with Kirchhoff's laws for electric circuits, in which case similar conditions apply to currents and voltages though/over electric components in a network.

Substructure Connectivity

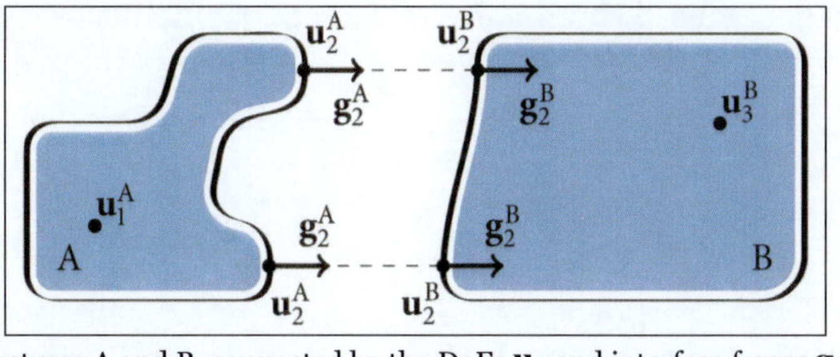

Assembly of two substructures A and B, connected by the DoFs $\mathbf{u}_2$ and interface forces $\mathbf{g}_2$ of the coupling nodes.

Consider two substructures A and B as depicted in the figure. The two substructures comprise a total of six nodes; the displacements of the nodes are described by a set of Degrees of Freedom (DoFs). The DoFs of the six nodes are partitioned as follows:

- DoFs of the internal nodes of substructure A;
- DoFs of the coupling nodes of substructures A and B, i.e. interface DoFs;
- DoFs of the internal nodes of substructure B.

Note that the denotation 1, 2 and 3 indicates the *function* of the nodes/DoFs rather than the total amount. Let us define the sets of DoFs for the two substructures A and B in concatenated form. The displacements and applied forces are represented by the sets $\mathbf{u}$ and $\mathbf{f}$. For the purpose of substructuring, a set of interface forces **g** is introduced which only contains non-zero entries on the interface DoFs:

$$u \triangleq \begin{bmatrix} u_1^A \\ u_2^A \\ u_2^B \\ u_3^B \end{bmatrix} \quad f \triangleq \begin{bmatrix} f_1^A \\ f_2^A \\ f_2^B \\ f_3^B \end{bmatrix} \quad g \triangleq \begin{bmatrix} 0 \\ g_2^A \\ g_2^B \\ 0 \end{bmatrix} \quad u, f, g \in \mathbb{R}^n$$

The relation between dynamic displacements u and applied forces f of the uncoupled problem is governed by a particular dynamic equation, The uncoupled equations of motion are augmented by extra terms/equations for compatibility and equilibrium.

Compatibility

The *compatibility condition* requires that the interface DoFs have the same sign and value at both sides of the interface: $\mathbf{u}_2^A = \mathbf{u}_2^B$. This condition can be expressed using a so-called *signed Boolean matrix*, denoted by **B**. For the given example this can be expressed as:

$$\mathbf{Bu} = \mathbf{0} \quad \Rightarrow \quad \mathbf{u}_2^B - \mathbf{u}_2^A = \mathbf{0} \quad \Rightarrow \quad \mathbf{B} \triangleq \begin{bmatrix} \mathbf{0} & -\mathbf{I} & \mathbf{I} & \mathbf{0} \end{bmatrix}$$

In some cases the interface nodes of the substructures are non-conforming, e.g. when two substructures are meshed separately. In such cases a non-Boolean matrix **B** has to be used in order to enforce a weak interface compatibility.

A second form in which the compatibility condition can be expressed is by means of coordinate substitution by a set of generalised coordinates **q**. The set **q** contains the unique coordinates that remain after assembly of the substructures Every matching pair of interface DoFs is described by a single generalised coordinate, which means that the compatibility condition is automatically enforced. Expressing **u** using **q** gives:

$$\mathbf{u} = \mathbf{Lq} \Rightarrow \begin{cases} \mathbf{u}_1^A = \mathbf{q}_1 \\ \mathbf{u}_2^A = \mathbf{q}_2 \\ \mathbf{u}_2^B = \mathbf{q}_2 \\ \mathbf{u}_3^B = \mathbf{q}_3 \end{cases} \Rightarrow \quad \mathbf{L} \triangleq \begin{bmatrix} \mathbf{I} & \mathbf{0} & \mathbf{0} \\ \mathbf{0} & \mathbf{I} & \mathbf{0} \\ \mathbf{0} & \mathbf{I} & \mathbf{0} \\ \mathbf{0} & \mathbf{0} & \mathbf{I} \end{bmatrix}$$

Matrix **L** is referred to as the *Boolean localisation matrix*. A useful relation between matrix **B** and **L** can be exposed by noting that compatibility should hold for any set of physical coordinates **u** expressed by **q**. Indeed, substituting $\mathbf{u} = \mathbf{Lq}$ in the equation $\mathbf{Bu} = \mathbf{0}$:

$$\mathbf{Bu} = \mathbf{BLq} = \mathbf{0} \quad \forall \mathbf{q}$$

Hence **L** represents the nullspace of **B**:

$$\begin{cases} \mathbf{L} = \text{null}(\mathbf{B}) \\ \mathbf{B}^T = \text{null}(\mathbf{L}^T) \end{cases}$$

This means in practice that one only needs to define **B** or **L**; the other Boolean matrix is calculated using the nullspace property.

Equilibrium

The second condition that has to be satisfied for substructure assembly is the *force equilibrium* for matching interface forces **g**. For the current example, this condition can be written as $\mathbf{g}_2^A = -\mathbf{g}_2^B$. Similar to the compatibility equation, the force equilibrium condition can be expressed using a Boolean matrix. Use is made of the transpose of the Boolean localisation matrix **L** that was introduced to write compatibility:

$$\mathbf{L}^T\mathbf{g} = \mathbf{0} \Rightarrow \begin{cases} \mathbf{g}_1^A = \mathbf{0} \\ \mathbf{g}_2^A + \mathbf{g}_2^B = \mathbf{0} \\ \mathbf{g}_3^B = \mathbf{0} \end{cases}$$

The equations for $\mathbf{g}_1$ and $\mathbf{g}_3$ state that the interface forces on internal nodes are zero, hence not present. The equation for $\mathbf{g}_2$ correctly establishes the force equilibrium between a matching pair of interface DoFs according to Newton's third law.

A second notation in which the equilibrium condition can be expressed is by introducing a set of Lagrange multipliers λ. The substitution of these Lagrange multipliers is possible as $\mathbf{g}_2^A$ and $\mathbf{g}_2^B$ differ only in sign, not in value. Using again the signed Boolean matrix **B**:

$$\mathbf{g} = -\mathbf{B}^T\lambda \Rightarrow \begin{cases} \mathbf{g}_1^A = \mathbf{0} \\ \mathbf{g}_2^A = \lambda \\ \mathbf{g}_2^B = -\lambda \\ \mathbf{g}_3^B = \mathbf{0} \end{cases}$$

The set λ defines the intensity of the interface forces $\mathbf{g}_2$. Each Lagrange multiplier represents the magnitude of two matching interface forces in the assembly. By defining the interface forces **g** using Lagrange multipliers λ, force equilibrium is automatically satisfied. This can be seen by substituting $\mathbf{g} = -\mathbf{B}^T\lambda$ into the first equilibrium equation:

$$\mathbf{L}^T\mathbf{g} = -\mathbf{L}^T\mathbf{B}^T\lambda = \mathbf{0} \quad \forall\mathbf{g}$$

Again, the nullspace property of the Boolean matrices is used here, namely: $\mathbf{L}^T\mathbf{B}^T = \mathbf{0}$.

The two conditions as presented above can be applied to establish coupling / decoupling in a myriad of domains and are thus independent of variables such as time, frequency, mode, etc. Some implementations of the interface conditions for the most common domains of substructuring are presented below.

Substructuring in the Physical Domain

The physical domain is thc domain that has the most straightforward physical interpretation. For each discrete linearised dynamic system one is able to write an equilibrium between the externally applied forces and the internal forces originating from intrinsic inertia, viscous damping and elasticity. This relation is governed by one of the most elementary formulas in structural vibrations:

$$\mathbf{M}\ddot{\mathbf{u}}(t)+\mathbf{C}\dot{\mathbf{u}}(t)+\mathbf{K}\mathbf{u}(t)=\mathbf{f}(t)$$

$\mathbf{M},\mathbf{C},\mathbf{K}$ represent the mass, damping and stiffness matrix of the system. These matrices are often obtained from finite element modelling (FEM), and are referred to as the numerical model of the structure. Furthermore, $\mathbf{u}$ represents the DoFs and $\mathbf{f}$ the force vector which are dependent on time (t). This dependency is omitted in the following equations in order to improve readability.

Coupling in the Physical Domain

Coupling of n substructures in the physical domain first requires writing the uncoupled equations of motion of the n substructures in block diagonal form:

$$\mathbf{M}\triangleq \operatorname{diag}(\mathbf{M}^{(1)},\ldots,\mathbf{M}^{(n)})=\begin{bmatrix}\mathbf{M}^{(1)} & . & . \\ . & \ddots & . \\ . & . & \mathbf{M}^{(n)}\end{bmatrix}$$

$$\mathbf{C}\triangleq \operatorname{diag}(\mathbf{C}^{(1)},\ldots,\mathbf{C}^{(n)}) \quad \mathbf{K}\triangleq \operatorname{diag}(\mathbf{K}^{(1)},\ldots,\mathbf{K}^{(n)})$$

$$\mathbf{u}\triangleq\begin{bmatrix}\mathbf{u}^{(1)}\\ \vdots \\ \mathbf{u}^{(n)}\end{bmatrix},\quad \mathbf{f}\triangleq\begin{bmatrix}\mathbf{f}^{(1)}\\ \vdots \\ \mathbf{f}^{(n)}\end{bmatrix},\quad \mathbf{g}\triangleq\begin{bmatrix}\mathbf{g}^{(1)}\\ \vdots \\ \mathbf{g}^{(n)}\end{bmatrix}$$

Next, two assembly approaches can be distinguished: primal and dual assembly.

Primal Assembly

For primal assembly, a unique set of degrees of freedom $\mathbf{q}$ is defined in order to satisfy compatibility, $\mathbf{u} \quad \mathbf{Lq}$. Furthermore, a second equation is added to enforce interface force equilibrium. This results in the following coupled dynamic equilibrium equations:

$$\begin{cases}\mathbf{ML}\ddot{\mathbf{q}}+\mathbf{CL}\dot{\mathbf{q}}+\mathbf{KLq}=\mathbf{f}+\mathbf{g}\\ \mathbf{L}^{\mathrm{T}}\mathbf{g}=\mathbf{0}\end{cases}$$

Pre-multiplying the first equation by $\mathbf{L}^{T}$ and noting that $\mathbf{L}^{\mathrm{T}}\mathbf{g}=\mathbf{0}$, the primal assembly reduces to:

$$\tilde{\mathbf{M}}\ddot{\mathbf{q}}+\tilde{\mathbf{C}}\dot{\mathbf{q}}+\tilde{\mathbf{K}}\mathbf{q}=\mathbf{L}^{\mathrm{T}}\mathbf{f} \quad \begin{cases}\tilde{\mathbf{M}}=\mathbf{L}^{\mathrm{T}}\mathbf{ML}\\ \tilde{\mathbf{C}}=\mathbf{L}^{\mathrm{T}}\mathbf{CL}\\ \tilde{\mathbf{K}}=\mathbf{L}^{\mathrm{T}}\mathbf{KL}\end{cases}$$

The primally assembled system matrices can be used for a transient simulation by any standard time stepping algorithm. Note that the primal assembly technique is analogue to assembly of super-elements in finite element methods.

Dual Assembly

In the dual assembly formulation the global set of DoFs is retained and an assembly is made by a priori satisfying the equilibrium condition $\mathbf{g} = -\mathbf{B}^{\mathrm{T}}\lambda$. Again, the Lagrange multipliers represent the interface forces connecting the DoFs at the interface. As these are unknowns, they are moved to the left-hand side of the equation. In order to satisfy compatibility, a second equation is added to the system, now operating on the displacements:

$$\begin{cases} \mathbf{M}\ddot{\mathbf{u}} + \mathbf{C}\dot{\mathbf{u}} + \mathbf{K}\mathbf{u} + \mathbf{B}^{\mathrm{T}}\lambda = \mathbf{f} \\ \mathbf{B}\mathbf{u} = \mathbf{0} \end{cases}$$

The dually assembled system can be written in matrix form as:

$$\begin{bmatrix} \mathbf{M} & \mathbf{0} \\ \mathbf{0} & \mathbf{0} \end{bmatrix} \begin{bmatrix} \ddot{\mathbf{u}} \\ \lambda \end{bmatrix} + \begin{bmatrix} \mathbf{C} & \mathbf{0} \\ \mathbf{0} & \mathbf{0} \end{bmatrix} \begin{bmatrix} \dot{\mathbf{u}} \\ \lambda \end{bmatrix} + \begin{bmatrix} \mathbf{K} & \mathbf{B}^{T} \\ \mathbf{B} & \mathbf{0} \end{bmatrix} \begin{bmatrix} \mathbf{u} \\ \lambda \end{bmatrix} = \begin{bmatrix} \mathbf{f} \\ \mathbf{0} \end{bmatrix}$$

This dually assembled system can also be used in a transient simulation by means of a standard time stepping algorithm.

Substructuring in the Frequency Domain

In order to write out the equations for frequency based substructuring (FBS), the dynamic equilibrium first has to be put in the frequency domain. Starting with the dynamic equilibrium in the physical domain:

$$\mathbf{M}\ddot{\mathbf{u}}(t) + \mathbf{C}\dot{\mathbf{u}}(t) + \mathbf{K}\mathbf{u}(t) = \mathbf{f}(t)$$

Taking the Fourier transform of this equation gives the dynamic equilibrium in the frequency domain:

$$\mathbf{Z}(\omega)\mathbf{u}(\omega) = \mathbf{f}(\omega) \quad \text{with} \quad \mathbf{Z}(\omega) = \left[-\omega^2\mathbf{M} + j\omega\mathbf{C} + \mathbf{K}\right]$$

Matrix $\mathbf{Z}(\omega)$ is referred to as the dynamic stiffness matrix. This matrix consists of the complex-valued frequency-dependent functions that describe the force required to generate a unit harmonic displacement at a certain DoF. The inverse of the matrix $\mathbf{Z}(\omega)$ is defined as $\mathbf{Y}(\omega) \triangleq (\mathbf{Z}(\omega))^{-1}$ and yields the more intuitive admittance notation:

$$\mathbf{u}(\omega) = \mathbf{Y}(\omega)\mathbf{f}(\omega)$$

The receptance matrix $\mathbf{Y}(\omega)$ contains the frequency response functions (FRFs) of the structure which describe the displacement response to a unit input force. Other variants of the receptance matrix are the mobility and accelerance matrix, which respectively describe the velocity and acceleration response. The elements of the dynamic stiffness (or *impedance* in general) and receptance (or *admittance* in general) matrix are defined as follows:

$$\text{Impedance:} \quad Z_{ij}(\omega) = \frac{f_i(\omega)}{u_j(\omega)} \qquad u_{k \neq j} = 0$$

$$\text{Admittance:} \quad Y_{ij}(\omega) = \frac{u_i(\omega)}{f_j(\omega)} \qquad f_{k \neq j} = 0$$

Coupling in the Frequency Domain

In order to couple two substructures in the frequency domain, use is made of the admittance and impedance matrices of both substructures. Using the definition of substructures A and B as introduced previously, the following impedance and admittance matrices are defined (note that the frequency dependency (ω) is omitted from the terms to improve readability):

$$\mathbf{Z}^A \triangleq \begin{bmatrix} \mathbf{Z}_{11}^A & \mathbf{Z}_{12}^A \\ \mathbf{Z}_{21}^A & \mathbf{Z}_{22}^A \end{bmatrix} \quad \mathbf{Z}^B \triangleq \begin{bmatrix} \mathbf{Z}_{22}^B & \mathbf{Z}_{23}^B \\ \mathbf{Z}_{32}^B & \mathbf{Z}_{33}^B \end{bmatrix}$$

$$\mathbf{Y}^A \triangleq \begin{bmatrix} \mathbf{Y}_{11}^A & \mathbf{Y}_{12}^A \\ \mathbf{Y}_{21}^A & \mathbf{Y}_{22}^A \end{bmatrix} \quad \mathbf{Y}^B \triangleq \begin{bmatrix} \mathbf{Y}_{22}^B & \mathbf{Y}_{23}^B \\ \mathbf{Y}_{32}^B & \mathbf{Y}_{33}^B \end{bmatrix}$$

The two admittance and impedance matrices can be put in block diagonal form in order to align with the global set of DoFs $\mathbf{u}$:

$$\mathbf{Z} \triangleq \begin{bmatrix} \mathbf{Z}^A & \mathbf{0} \\ \mathbf{0} & \mathbf{Z}^B \end{bmatrix} \triangleq \begin{bmatrix} \mathbf{Z}_{11}^A & \mathbf{Z}_{12}^A & \mathbf{0} & \mathbf{0} \\ \mathbf{Z}_{21}^A & \mathbf{Z}_{22}^A & \mathbf{0} & \mathbf{0} \\ \mathbf{0} & \mathbf{0} & \mathbf{Z}_{22}^B & \mathbf{Z}_{23}^B \\ \mathbf{0} & \mathbf{0} & \mathbf{Z}_{32}^B & \mathbf{Z}_{33}^B \end{bmatrix}$$

$$\mathbf{Y} \triangleq \begin{bmatrix} \mathbf{Y}^A & \mathbf{0} \\ \mathbf{0} & \mathbf{Y}^B \end{bmatrix} \triangleq \begin{bmatrix} \mathbf{Y}_{11}^A & \mathbf{Y}_{12}^A & \mathbf{0} & \mathbf{0} \\ \mathbf{Y}_{21}^A & \mathbf{Y}_{22}^A & \mathbf{0} & \mathbf{0} \\ \mathbf{0} & \mathbf{0} & \mathbf{Y}_{22}^B & \mathbf{Y}_{23}^B \\ \mathbf{0} & \mathbf{0} & \mathbf{Y}_{32}^B & \mathbf{Y}_{33}^B \end{bmatrix}$$

The off-diagonal zero terms show that at this point no coupling is present between the two substructures. In order to create this coupling, use can be made of the primal or dual assembly method. Both assembly methods make use of the dynamic equations as was defined before:

$$\mathbf{Zu} = \mathbf{f} + \mathbf{g}$$
$$\mathbf{u} = \mathbf{Y}(\mathbf{f} + \mathbf{g})$$

In these equations $\mathbf{g}$ is again used to define the set of interface forces, which are yet unknown.

Primal Assembly

In order to obtain the primal system of equations, a unique set of coordinates $\mathbf{q}$ is defined: $\mathbf{u} = \mathbf{Lq}$. By definition of an appropriate Boolean localisation matrix $\mathbf{L}$, a unique set of DoFs remains for which the compatibility condition is satisfied a priori (*compatibility condition*). In order to satisfy the *equilibrium condition* a second equation is added to the equations of motion:

$$\begin{cases} \mathbf{ZLq} = \mathbf{f} + \mathbf{g} \\ \mathbf{L}^T\mathbf{g} = \mathbf{0} \end{cases}$$

Pre-multiplying the first equation with $\mathbf{L}^T$ yields the notation of the assembled equations of motion for the generalised coordinates $\mathbf{q}$:

$$\tilde{\mathbf{Z}}\mathbf{q} = \tilde{\mathbf{f}} \quad \text{with} \quad \begin{cases} \tilde{\mathbf{Z}} = \mathbf{L}^T\mathbf{Z}\mathbf{L} \\ \tilde{\mathbf{f}} = \mathbf{L}^T\mathbf{f} \end{cases}$$

This result can be rewritten in admittance form as:

$$\mathbf{q} = \tilde{\mathbf{Y}}\tilde{\mathbf{f}} \quad \text{with} \quad \tilde{\mathbf{Y}} = (\tilde{\mathbf{Z}})^{-1}$$

This last result gives access to the generalised responses as a result of the generalised applied forces $\tilde{\mathbf{f}}$, namely by inverting the primally assembled impedance matrix.

The primal assembly procedure is mainly of interest when one has access to the dynamics in impedance form, e.g. from finite element modelling. When one only has access to the dynamics in admittance notation, the dual formulation is a more suitable approach.

Dual Assembly

A dually assembled system starts with the system written in the admittance notation. For a dually assembled system the force equilibrium condition is satisfied a priori by substituting Lagrange multipliers λ for the interface forces: $\mathbf{g} = -\mathbf{B}^T\lambda$. The compatibility condition is enforced by adding an additional equation:

$$\begin{cases} \mathbf{u} = \mathbf{Y}(\mathbf{f} - \mathbf{B}^T\lambda) \\ \mathbf{B}\mathbf{u} = \mathbf{0} \end{cases}$$

Substituting the first line in the second and solving for λ gives:

$$\lambda = (\mathbf{B}\mathbf{Y}\mathbf{B}^T)^{-1}\mathbf{B}\mathbf{Y}\mathbf{f}$$

The term $\mathbf{BYf}$ represents the incompatibility caused by the uncoupled responses of the substructures to the applied forces $\mathbf{f}$. By multiplying the incompatibility with the combined interface stiffness, i.e. $(\mathbf{B}\mathbf{Y}\mathbf{B}^T)^{-1}$, the forces λ that keep the substructures together are determined. The coupled response is obtained by substituting the calculated λ back into the original equation:

$$\mathbf{u} = \mathbf{Y}\mathbf{f} - \mathbf{Y}\mathbf{B}^T(\mathbf{B}\mathbf{Y}\mathbf{B}^{\mathrm{T}})^{-1}\mathbf{B}\mathbf{Y}\mathbf{f}$$

$$\mathbf{u} = \tilde{\mathbf{Y}}\mathbf{f} \quad \text{with:} \quad \tilde{\mathbf{Y}} = \left(\mathbf{I} - \mathbf{Y}\mathbf{B}^T(\mathbf{B}\mathbf{Y}\mathbf{B}^{\mathrm{T}})^{-1}\mathbf{B}\right)\mathbf{Y}$$

This coupling method is referred to as the Lagrange-multiplier frequency-based substructuring (LM-FBS) method. The LM-FBS method allows for quick and easy assembling of an arbitrary number of substructures in a systematic fashion. Note that the result $\tilde{\mathbf{Y}}$ is theoretically the same as was obtained above by application of primal assembly.

Decoupling in the Frequency Domain

In addition to coupling of substructures, one is also able to decouple substructures from assemblies. Using the plus sign as a substructure coupling operator, the coupling procedure could simply be described as AB = A + B. Using a similar notation, decoupling could be formulated as AB - B = A.

Decoupling procedures are often required to remove substructures that were added for measurement purposes, e.g. to fix the structure. Similar to coupling, a primal and dual formulation exists for decoupling procedures.

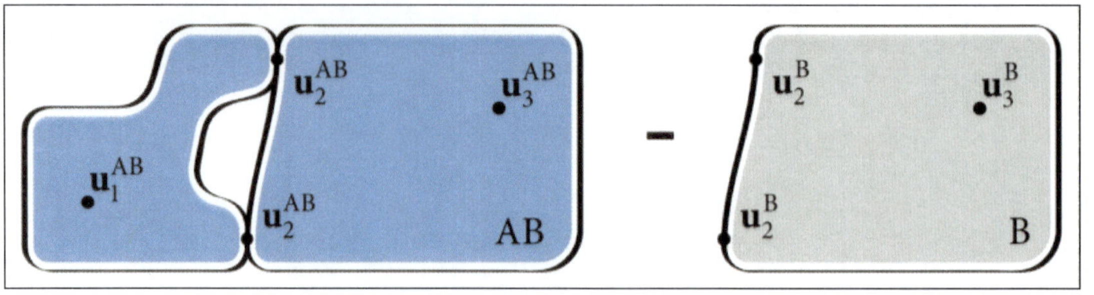

Decoupling of substructure B from assembly AB.

Primal Disassembly

As a result of the primal coupling, the impedance matrix of the assembled system can be written as follows:

$$\mathbf{Z}^{AB} = \begin{bmatrix} \mathbf{Z}_{11}^{AB} & \mathbf{Z}_{12}^{AB} & \mathbf{Z}_{13}^{AB} \\ \mathbf{Z}_{21}^{AB} & \mathbf{Z}_{22}^{AB} & \mathbf{Z}_{23}^{AB} \\ \mathbf{Z}_{31}^{AB} & \mathbf{Z}_{32}^{AB} & \mathbf{Z}_{33}^{AB} \end{bmatrix} = \begin{bmatrix} \mathbf{Z}_{11}^{A} & \mathbf{Z}_{12}^{A} & \mathbf{0} \\ \mathbf{Z}_{21}^{A} & \mathbf{Z}_{22}^{A} + \mathbf{Z}_{22}^{B} & \mathbf{Z}_{23}^{B} \\ \mathbf{0} & \mathbf{Z}_{32}^{B} & \mathbf{Z}_{33}^{B} \end{bmatrix}$$

Using this relation, the following trivial subtraction operation would suffice for the decoupling of the substructure B from assembly AB:

$$\begin{bmatrix} \mathbf{Z}_{11}^{A} & \mathbf{Z}_{12}^{A} & \mathbf{0} \\ \mathbf{Z}_{21}^{A} & \mathbf{Z}_{22}^{A} + \mathbf{Z}_{22}^{B} & \mathbf{Z}_{23}^{B} \\ \mathbf{0} & \mathbf{Z}_{32}^{B} & \mathbf{Z}_{33}^{B} \end{bmatrix} - \begin{bmatrix} \mathbf{0} & \mathbf{0} & \mathbf{0} \\ \mathbf{0} & \mathbf{Z}_{22}^{B} & \mathbf{Z}_{23}^{B} \\ \mathbf{0} & \mathbf{Z}_{32}^{B} & \mathbf{Z}_{33}^{B} \end{bmatrix} = \begin{bmatrix} \mathbf{Z}_{11}^{A} & \mathbf{Z}_{12}^{A} & \mathbf{0} \\ \mathbf{Z}_{21}^{A} & \mathbf{Z}_{22}^{A} & \mathbf{0} \\ \mathbf{0} & \mathbf{0} & \mathbf{0} \end{bmatrix}$$

By placing the impedance of AB and B in block-diagonal form, with a minus sign for the impedance of B to account for the subtraction operation, the same equation that was used for primal coupling can now be used to perform the primal decoupling procedures.

$$\mathbf{Z} \triangleq \begin{bmatrix} \mathbf{Z}^{AB} & \mathbf{0} \\ \mathbf{0} & -\mathbf{Z}^{B} \end{bmatrix} \quad \Rightarrow \quad \tilde{\mathbf{Z}} = \mathbf{L}^{T}\mathbf{Z}\mathbf{L}$$

with:

$$\mathbf{L} = \begin{bmatrix} \mathbf{I} & \mathbf{0} & \mathbf{0} \\ \mathbf{0} & \mathbf{I} & \mathbf{0} \\ \mathbf{0} & \mathbf{0} & \mathbf{I} \\ \mathbf{0} & \mathbf{I} & \mathbf{0} \\ \mathbf{0} & \mathbf{0} & \mathbf{I} \end{bmatrix}$$

The primal disassembly can thus be understood as the assembly of structure AB with the negative impedance of substructure B. A limitation of the primal disassembly is that all DoF of the substructure that is to be decoupled have to be exactly represented in the assembled situation. For numerical decoupling situations this should not pose any problems, however for experimental cases this can be troublesome. A solution to this problem can be found in the dual disassembly.

Dual Disassembly

Similar to the dual assembly, the dual disassembly approaches the decoupling problem using the admittance matrices. Decoupling in the dual domain means finding a force that ensures compatibility, yet acts in the opposite direction. This newly found force would then counteract the force that is applied to the assembly due to the dynamics of substructure B. Writing this out in equations of motion:

$$\begin{cases} \mathbf{u}^{AB} = \mathbf{Y}^{AB}\mathbf{f} + \mathbf{Y}^{AB}\mathbf{g} \\ \mathbf{u}^{B} = -\mathbf{Y}^{B}\mathbf{g} \end{cases}$$

In order to write the dynamics of both systems in one equation, using the LM-FBS assembly notation, the following matrices are defined:

$$\mathbf{Y} \triangleq \begin{bmatrix} \mathbf{Y}^{AB} & \mathbf{0} \\ \mathbf{0} & -\mathbf{Y}^{B} \end{bmatrix}$$

$$\mathbf{u} = \begin{bmatrix} \mathbf{u}_1^{AB} \\ \mathbf{u}_2^{AB} \\ \mathbf{u}_3^{AB} \\ \mathbf{u}_2^{B} \\ \mathbf{u}_3^{B} \end{bmatrix}; \quad \mathbf{f} \triangleq \begin{bmatrix} \mathbf{f}_1^{AB} \\ \mathbf{f}_2^{AB} \\ \mathbf{0} \\ \mathbf{0} \\ \mathbf{0} \end{bmatrix}; \quad \mathbf{g} \triangleq \begin{bmatrix} \mathbf{0} \\ \mathbf{g}_2^{AB} \\ \mathbf{0} \\ \mathbf{g}_2^{B} \\ \mathbf{0} \end{bmatrix}$$

In order to enforce compatibility, a similar approach is used as for the assembly task. Defining a **B** -matrix to enforce compatibility:

$$\mathbf{B} = \begin{bmatrix} \mathbf{0} & -\mathbf{I} & \mathbf{0} & \mathbf{I} & \mathbf{0} \end{bmatrix}$$

Using this notation, the disassembly procedure can be performed using exactly the same equation as was used for the dual assembly:

$$\begin{cases} \mathbf{u} = \mathbf{Y}(\mathbf{f} - \mathbf{B}^{T}\lambda) \\ \mathbf{B}\mathbf{u} = \mathbf{0} \end{cases}$$

This means that coupling and decoupling procedures using LM-FBS require identical steps, the only difference being the manner in which the global admittance matrix is defined. Indeed, the substructures to couple appear with a plus sign, whereas decoupled structures carry a minus sign:

$$\text{Coupling} \quad \mathbf{Y} \triangleq \begin{bmatrix} \mathbf{Y}_A & \mathbf{0} \\ \mathbf{0} & \mathbf{Y}_B \end{bmatrix}$$

$$\text{Decoupling} \quad \mathbf{Y} \triangleq \begin{bmatrix} \mathbf{Y}_{AB} & \mathbf{0} \\ \mathbf{0} & -\mathbf{Y}_B \end{bmatrix}$$

More advanced decoupling techniques use the fact that internal points of substructure B appear in both the admittances of AB and B, hence can be used to enhance the decoupling process.

Vehicle Dynamics

As a basic theory of the vehicle industry, the vehicle dynamics plays an important role in the development of the vehicle industry. In the past decades, great progress was made in the theory and experiment of vehicle dynamics. In vehicle dynamics, the vehicle body (sprung mass), the suspension component (spring and damper) and tire (unsprung mass) are essential parts of the system.

The vehicle dynamics plays an important role in the development of the vehicle industry. In the past decades, great progress was made in the theory and experiment of vehicle dynamics. In vehicle dynamics, the vehicle body (sprung mass), the suspension component (spring and damper) and tire (unsprung mass) are essential parts of the system.

The global vehicle industry and market structure experienced an unprecedented scale of change in the 1990s. There has been an increasing demand on vehicle safety, environmental protection and intelligent control. Thus, the advanced technologies such as computer technology, virtual reality technology and intelligent algorithm have been widely used in the vehicle industry. The vehicle dynamics plays an important role in the development of vehicle industry. The early research of vehicle dynamics concentrated on various working conditions and service performances under external excitation. In the 1930s, researchers began to focus on the steering, suspension mechanics and driving stability. Lanchester Maurice and Segel studied the effects of the external environment (such as road surface roughness, air flow, tire and driver) on the vehicle dynamics and the coupling interaction of these conditions. In 1993 Segel presented a comprehensive summary on the achievements of the vehicle dynamics before 1990 in the Proceedings of Institution of Mechanical Engineers. In the following decades, the vehicle ride comfort and handing stability research have been widely investigated. The handling dynamics deals with the lateral dynamics or transverse dynamics of the vehicle, which mainly refer to vehicle handling stability, vehicle sideslip caused by tire lateral force, yawing and roll motion. The handing stability of vehicle dynamics research went through the development from experimental studies to the theoretical analysis, from the open-loop to the closed-loop. The representative monographs of vehicle handling dynamics include "Vehicle Handling Dynamics Theory and Application" by Abe, "Vehicle Handling Dynamics Theory" written by Guo. The vehicle driving dynamics is divided into longitudinal dynamics and vertical dynamics, which includes driving, braking and ride comfort. The problem of driving slip and braking slip are solved by the study of vehicle longitudinal tire force, which can also improve driving and braking efficiency. The ride comfort focuses on vehicle vibration and pitch movement caused by vertical tire force. The representative monographs are "Vehicle Dynamics and Control" written by Rajamani, "Vehicle Dynamics Theory and Applications" written by Zhang. In addition, the field of vehicle dynamics research also involves the longitudinal force of tire when vehicles are speeding up or braking and vehicle vibration caused by the working engine, etc.

In vehicle dynamics, the vehicle body (sprung mass), the suspension component (spring and damper) and tire (unsprung mass) are essential parts of the system. The vehicle-road coupling is another important aspect of vehicle dynamics.

Vehicle Dynamics Model

The vehicle dynamics models went through the development from the traditional lumped parameter model to the finite element model (FEM), the dynamical substructure model and the multibody system dynamics model, from the linear model to the non-linear model with the non-linear stiffness and the non-linear damping.

Lumped Parameter Model

For the lumped parameter modeling method, the finite degree of freedom (DOF) model of a vehicle system is comprised of mass, spring and damper elements. The examples include a quarter vehicle model with 2 DOF, half vehicle model with four or five DOF, full vehicle model with 7 or 18 DOF, etc.

For the vehicle handling stability research, the number of DOF can be two, ten or more such as 2 DOF (lateral, yaw), 3 DOF (longitudinal, lateral, yaw) and 4 DOF (longitudinal, lateral, yaw, roll). It is known that a 2 DOF vehicle model is usually applied when the vehicle lateral acceleration is less than 0.3 g. If the tire cornering properties is in the linear range, the vehicle model can be simplified as a 3 DOF model with lateral, longitudinal and yaw movement. The model is also widely used to study vehicle handling dynamics while the vehicle encounters emergency collision avoidance. Liu proposed a four-wheel-steering vehicle nonlinear lateral dynamics model of 3 DOF based on Pacejka model. The experimental study and simulation analysis for the vehicle single lane change, double lane change are also complete. Huh proposed a six-axis heavy vehicle model of 18 DOF with MATLAB/Simulink, in which the suspension and tires characteristics are nonlinear. It is found that the effects of middle steering axle on the yawing angular velocity and lateral acceleration cannot be ignored. The simulation research of handling stability went through the development from the simulation of the steady state response characteristics to the simulation of transient response characteristics and cornering braking characteristics. Based on 250 references, Vlk summed up the truck handling stability during longitudinal traveling, lateral traveling and braking. Nagai, put forward an active control model to control the front wheel steering angle, which can improve the vehicle dynamic performances under the condition of steering, braking, changing lane and wind disturbance. Zhao, reviewed the development history of vehicle dynamics and the main research methods. Examples of vehicle handling dynamics and inverse dynamics are given.

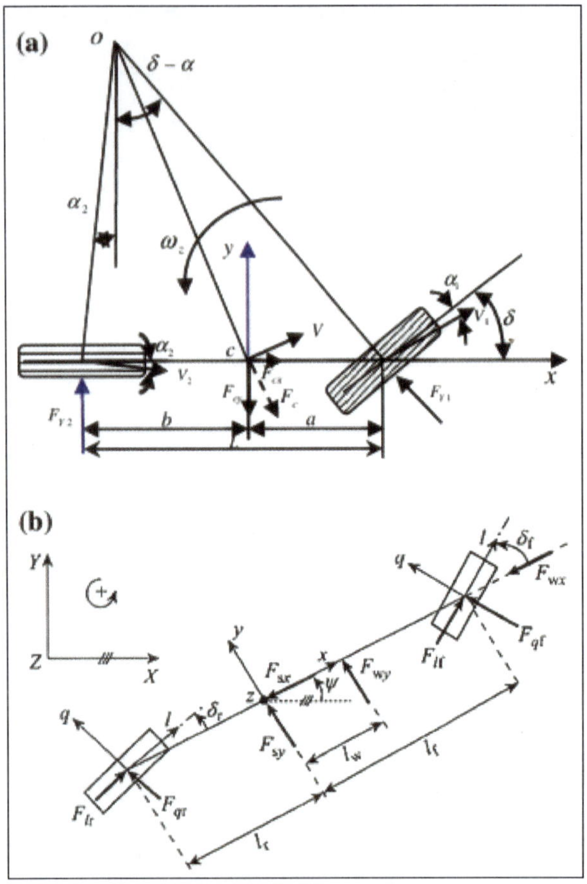

Lumped mass model of vehicle handling dynamics. a 2DOF (lateral, yaw) model, b 3 DOF(longitudinal, lateral, yaw) model.

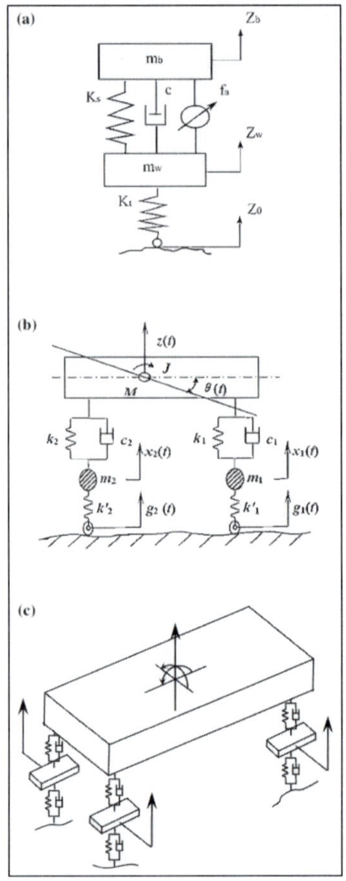

Lumped mass model for vehicle driving dynamics. a One-dimensional (quarter vehicle) model, b Two-dimensional (4 DOF) model, c Three-dimensional (7 DOF) full vehicle model.

For the vehicle driving dynamics, there is a quarter vehicle model of 2 DOF, which assumes the movement of the vehicle front axle and rear axle to vibrate independently. The simplest quarter vehicle model is one-dimensional model. It is widely applied in the research and design of the vehicle active and semi-active suspensions. The lumped mass vehicle model of 7, 8, or more DOF is usually three-dimensional, which can analyze vertical, pitch and roll vibrations of the vehicle. Cebon proposed a three-dimensional vehicle model with 14 DOF considering the longitudinal tire force, lateral tire force and the non-linear suspension. Several typical vehicle models for the driving dynamics are shown in figure below.

Although these lumped mass models are simplifications of the actual vehicle structure, they can capture the vehicle vibration characteristics and the effects of structural parameters from the vehicle performance. The advantage of these models lies in their simplicity and the possibility to obtain analytic descriptions and to design active or semi-active controls.

Virtual Prototyping Model

The finite element (FE) modeling demands much more calculation time. The FE vehicle model is mainly used for the dynamic design of complex parts such as vehicle body, engine mount, etc.

The dynamical substructure approach includes the mechanical admittance method and the modal synthesis method. The modal synthesis method has been successfully applied to study the vehicle vibration noise mechanism and vibration transfer characteristics of the vehicle chassis and frame. At present, the dynamical analysis modeling of the vehicle is usually combined with the lumped mass method, FE method and modal synthesis method.

With the development of computer technology and application software, the functional virtual prototyping (FVP) technology makes more and more complex models of the vehicle. The core contents of the FVP are the Multi-Body System Kinematics and Dynamics modeling theory. In recent years, some scholars begin to build the vehicle model and analyze its dynamic behavior by using the FVP software. The representative multi-body dynamics software includes Simpack, ADAMS and TruckSim. Valášek, established a virtual prototype of triaxial heavy vehicle model in Simpack software and designed the semi-active controller in MATLAB/Simulink. Hou and Ieluzzi developed a vehicle model in ADAMS/CAR and studied the semi-active suspension effectiveness for improving the vehicle vibration performance. Odhams developed a complex multi-body model of heavy load hinged vehicle using TruckSim software and studied the driving safety of trailer vehicle under the operation of active steering system. Ren, Yang and Lu investigated the dynamic behavior such as vehicle ride comfort, handling stability and so on, using ADAMS software. The most significant advantage of the multi-body vehicle model is the parametric modeling. In order to make it clear which vehicle parameter (vehicle geometric position, suspension stiffness and damping characteristics) has the most important effect on the vehicle performance, several trial simulations of the vehicle must be done. The traditional computational optimization approach becomes inadequate in dealing with the complex multi-body model of the vehicle with a lot non-linear parameters. A discrete optimization method proposed by Fisher, known as the design of experiment (DOE) for the optimization, has been adopted to study the vehicle design. This method has been widely used in the detection, earthquake prediction and product design due to its effectiveness, robustness, and high-quality solutions. Lu presented an orthogonal optimization program for a multi-body vehicle model. After several virtual experiments and range analysis, the most important influencing factor and its range are screened out. Then a matching scheme of vehicle parameters is proposed to achieve the ride comfort and the road friendliness optimization.

Although the lumped mass models of a vehicle can be used for the vehicle performance analysis and control study, they are too simple to show the nonlinear behavior of the suspension system and describe the detail of the vehicle structure. The virtual prototyping model (FE or multi-body) has advantages such as user-oriented interactive environment and virtual prototyping pretreatment function, which are important to the development of vehicle dynamics.

Tire Model

The tire is the link between a vehicle and the road surface. It does not only support the vehicle load, but also attenuates the shock from the uneven road surface. Thus, the tire dynamics plays an important role in the vehicle performance.

Tire Mechanical Model

The research about the vehicle tire mechanics originates from the aero tire research, which is dated back to the early 1930s. The tire mechanical model can be divided into three categories: pure physical model, empirical model and semi-empirical model.

According to the tire physical structure and working mechanism, the pure physical model is proposed in the form of mathematical expression to describe a tire mechanical characteristic. The pure physical models include Fiala, Gim and Dugoff tire model, etc. In the Fiala model, the belt and the buffer layers of a tire are simplified to an elastic beam under the lateral concentrated force. The tire lateral force and aligning torque is described by the dimensionless expressions. Although the calculation error of the Fiala model aligning torque is large, the cornering force has a better accuracy within 5oo slip angle, figure The Gim model is based on the Bergman 3D spring model, in which the tire side slipping and longitudinal sliding is governed by a kinetic equation of the infinitesimal element between the tire and road surface. It is widely used in the field of vehicle dynamics simulation and control study. The Dugoff model expresses the friction factor between the tire and the road as a function of the friction reduction factor and the friction when both the tread sliding speed and vehicle sliding velocity are zero. This gives a relationship between the tire-road surface friction coefficient and the cornering or longitudinal force in the Dugoff model. The tire can have large deformation under the loading, which is difficult to be described analytically. Therefore the pure physical model can only serve as the basis for the semi-empirical model.

From experimental data of tire, an empirical model can be directly built. The semi-empirical model has higher precision combined with the theoretical model and the experimental data. The famous Magic Formula model is proposed by Pacejka based on extensive experiments. It is a semi-empirical and semi-theoretical model, which is based on the tire physical prototype and the experimental data. It can be used to describe the six-directional wheel force in the steady state. In order to extend the frequency range of the tire, Pacekja introduced a dynamical rigid ring model based on the Magic Formula model and proposed a Swift model, figure The Magic Formula and Swift tire models were developed at Delft University of Technology. They were implemented in the commercial software (ADAMS, SIMPACK and MATLAB/Simulink), and named as MF-Tyre and MF-Swift, respectively. The mathematical expression of the tire lateral and longitudinal instantaneous deformation under any movement is calculated by Guo and Lu with a theoretical tire model in non-steady state. The calculation error of a pure theoretical model caused by unavoidable simplification can be compensated by the experiment and the data fitting in a semi-empirical model. Guo also put forward a semi-empirical model called UniTire Model that can describe a tire mechanical characteristics in an E exponential form. It describes the six-component characteristics of a tire in a variety of working conditions, and has high precision simulation for the complex and extreme conditions.

The FE model of the tire has experienced from the static and dynamic analysis to the thermal coupling, abrasion, fatigue life prediction and vehicle-tire- pavement coupled analysis. Narasimha used the ABAQUS software to investigate the lateral and longitudinal forces of pneumatic tires during steering and braking. Gall and Tkacik built a 3D FE model of tire with the tread, and found out that the normal stress reaches the maximum value when the friction coefficient is in the vicinity of 0.55. The distribution of normal stress tends to be stable with the increasing of friction coefficient. It is suited for the analysis of transient rolling contact, internal stress, modality, noise and so on. However, there are still limitations for the FE method. For example, the quantitative relationship between variable parameters cannot be expressed explicitly. Several typical tire models are shown in figure.

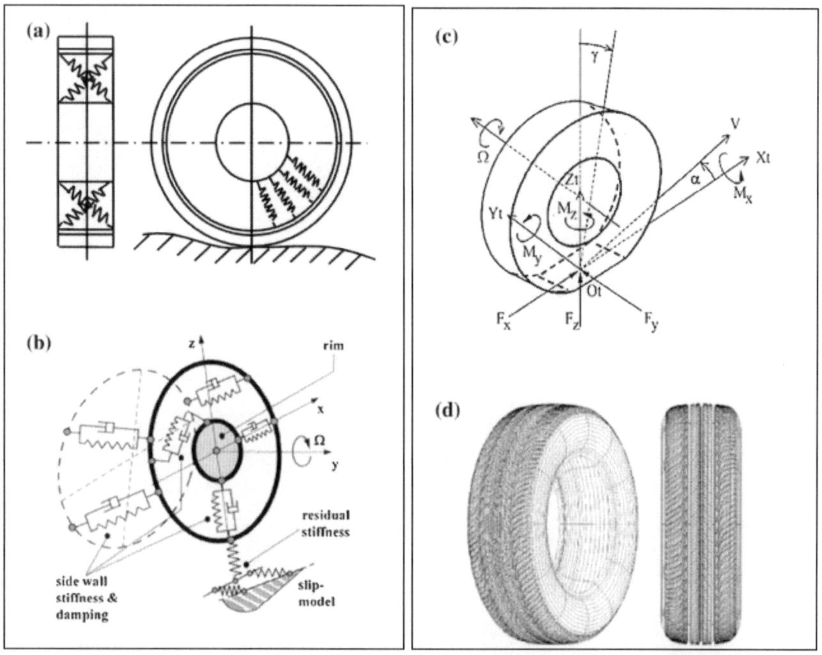

Tire models: (a) Physical model (Fiala), (b) Semi-empirical model (Swift), (c) Semi-empirical model (UniTire), (d) FE model.

Contact between Tire and Road

The relationship between the tire model and the road surface are usually assumed as the point contact. It is the convenient model for the dynamic analysis and widely used in the theoretical study of vehicle dynamics. Kyongsu established the road-surface roughness by using six sinusoidal waves and calculated the tire force by using the point contact model. The simulation result of the point contact tire model is rather similar to the measured tire force. Collins believed that the point contact model is adequate for the use in the shimmy analysis of the tire yaw and structural-torsion shimmy modes of vibration. Arunas and Jonas compared the results of the point-contact model, the flexible ring model and the flexible band model based on the quarter vehicle model. Costanzi and Cebon also used the point-contact tire model to calculate the tire dynamical force and the road deformation, which were validated by field data.

However, the point contact model is only appropriate for the vertical loading condition, in which the range of road surface excitation frequency is between 0.1 and 0.2 Hz. When the pavement wavelength is greater than 3 m, the model cannot reflect the motions of shortwave (high frequency) excitations. The high-frequency filtering property of a tire is called the enveloping characteristics, which is initially studied by Lippmann. The enveloping characteristics means that a tire has a property to envelope the road profile irregularities and attenuate the high spatial frequency components. Other tire models include line contact model or surface contact model. The spring and damping elements along with the imprinting length are used to form a fixed imprinting tire model. The imprinting length is invariant and the confluence of the tire force is always in the direction of the wheel center, which also has an enveloping characteristic within the imprinting. With this model, the calculated value of vertical axle load under the high frequency excitation is lower than the actual value. Guo presented a flexible roller contact tire model and carried out simulations of a vehicle vibration system based on the rigid and the flexible roller contact tire model. Guan and

Dong put forward a vehicle system model with an enveloping tire model to study the vehicle active suspension performances. The vehicle load distribution on the road surface is assumed to be uniform and symmetrical with parabolic or trapezoidal distribution. The true distribution of the loading on local contact area on a moving vehicle is not symmetrical. A non-uniform distribution model is introduced by Guo. A modified elastic roller tire model is proposed by Yang. A two-dimensional vehicle-road-subgrade coupling system based on the modified tire model is developed. The improved elastic roller line contact model and point contact model of the vehicle-road coupling system are compared.

There has been research work on experimental modeling of a tire. The South Africa researchers of Council for Scientific and Industrial Research (CSIR) developed a vehicle-road surface pressure transducer array (VRSPTA), commonly known as the "3-D Stress Sensor". Groenendik and Ronald proposed the simplified mathematical formula of the tire contact stress distribution based on the tire test results of freely rolling tire. A measurement device was developed and used to measure the tire ground pressure distribution of light vehicles with different tread patterns in different conditions of the tire inflation pressure and axle loading. A rig was developed for the tire impact test, allowing a drop mass with a round indenter to hit the pressurized tires with different impact energies. Although the cost of the test is relatively high, it provides important input to the mathematical tire model and analysis of physical properties.

Driver Model

A number of models have been developed to describe the human driving behavior and evaluate the vehicle safety in braking and turning. An early driver model based on the cognitive model was established in 1938 by Gibson and Crooks. In the 1980s, the number of publications about the driver model had reached a peak. McRuer is one of the scholars who has a great influence on the driver model of control theory. A long article written by MacAdam expounded the driver's limited driving ability in detail. Driver's behavior is not only limited to the control of extremities and the brain, but also related to the interaction with the outside world. It is a very complex process of perceiving location and speed as well as the action time, which involves the work of different parts of the brain (identification of position, decision-making, action to complete the task). Kinecke and Nielsen summarized the driver's behavior characteristics from the point of view of dynamics and control. Driver's behavior is nonlinear, time varying, adaptive and random. Furthermore, driver's response has the property of foresight and hysteresis. The foresight is manifested as predicting the development of traffic situation and preventing or handling in advance. Hysteresis shows up as delay in response to the events. Ranney puts forward several cognitive models of driving behavior. But, the model parameters are hard to be obtained. Cacciabue summarized the various factors that contribute to modelling human behaviour in the specialized environment. Baron studied an optimal driver control. Torsten and von Stryk proposed a control system combining the driver and vehicle factors in the optimal control theory. The driver's transient choice of path and speed has an influence on the vehicle model directly. But this factor did not include driver's motivation and personal characteristics. Cheng and Fujioka used the hierarchical decision-making system and the fuzzy system to design an appropriate distance between vehicles. Michon proposed a cognitive model of driver's decision-making process, which can achieve quantitative measurement and provide three types of decisions: strategic, technical and operational. Yang gave a longitudinal driver model and a lateral driver model to study the mechanism of accident using the optimal control theory. Guo proposed a closed-loop control system of human-vehicle-road and established a driver preview follower model as well as a preview optimal curvature model. The research of driver behavior in China is in its early stage.

Vehicle and Road Surface Interaction

In 1987, Frybal stated that "the dynamic issues of vehicle-pavement interaction is a new branch of the science". Since this research field has emerged dealing with dynamics problems of vehicle, road and the interaction between them. So far, the research field has not been fully developed yet except for a few publications. It is well recognized that the vehicle can cause the damage to the pavement. In 1958 the American association of highway and transportation officials (AASHO) began to focus on the pavement damage caused by the vehicle traveling. Based on many road tests under the different vehicle loadings, AASHO obtained the famous "fourth power law". The law states that there is a fourth power function relation between the fatigue damage of the pavement and the vehicle static axle load. The fourth power law is important to the design of pavement structure. However, the law doesn't consider the dynamical loading effect of vehicle on the pavement. In the 1970s, West Germany highway department proposed the evaluation of "Road Stress Factor", which took into account the effect of dynamical axle load on the pavement. The evaluation pointes out that the damage is largely caused by the vehicle axle load (including the static axial load and dynamical axle load) and the contact area between the tire and the road surface. Canadian roads and transport association organized a joint commission to research the vehicle quality and size. They studied the interaction between the truck and the pavement or bridge.

The purpose is to study the effect of design, size and quality safety limit of the truck on the roads, bridges, security and transport policy. In 1987, the United States Congress adopted and launched the Strategic Highway Research Program (SHRP), focusing on the research of four fields: asphalt road, pavement performance, concrete structure and road transportation. In the early 1990s, TRL's transportation institute of the university of Cambridge, UK, studied on interactions between the vehicle and road surface to assess the driving safety, suspension structure design, failure mechanism of asphalt pavement, research and development of dynamical weighing equipment, and so on. At the University of California, Beckeley, the damage caused by the tire pressure and the axial load is studied theoretically and experimentally. Kyongsu's research showed that the dynamical tire force is much higher than the static tire force. Both of the theoretical analysis and the experimental results proved that the semi-active suspension using double linear disturbance decoupling observer can reduce the tire dynamical force and the damage on the pavement. The European organization for the economic co-operation and development (OECD) put forward the road traffic research project entitled "Dynamic Interaction between Vehicles and Infrastructure Experiment (DIVINE)" and investigated the interaction mechanism of vehicles, pavement and bridge. A total of 19 countries had participated and conducted experiments in the indoor accelerated pavement test facility (CAPTIF) in Canterbury of New Zealand, on the proving ground in Virtta, Finland.

With the modal superposition and integral transformation method, Seong-Min and McCullough studied the dynamical response of plates on the viscous Winkler foundation under the moving loads with varying amplitudes. Huang and Thambiratnam also analyzed the dynamical response of thin plates with the Winkler foundation under the moving point harmonic load. Kim, Chen and Huang and many others studied the effects of vehicle parameters on multilayer pavement's dynamics with the finite element method. Mamlouk and Brademeyer proposed a concept of vehicle-pavement interaction, which can be applied to weigh-in-motion, pavement design, and performance and vehicle regulation. Cebon concluded that the interaction of vehicle and road is a weak coupling system for two reasons: (1) the displacement of the pavement is 0–1mm, much smaller than the deflection of the tire and suspension (10–20mm); (2) the speed of elastic waves in the road profile is 100–600m/s, much greater than the vehicle running speed (10–50 m/s).

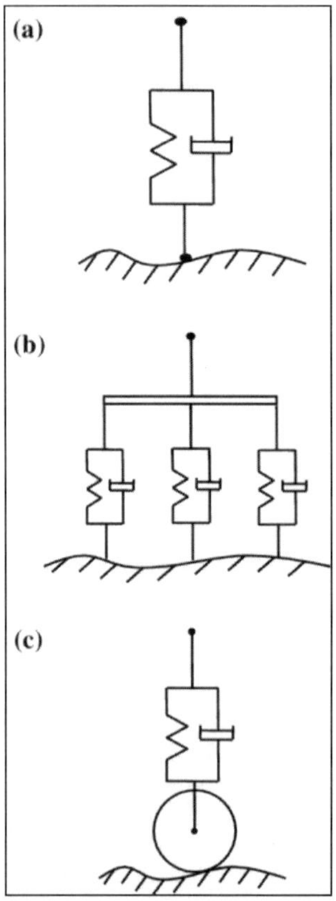

Contact relationship between tire and road surface: (a) Single point-contact model, (b) Fixed foot print model, (c) Enveloping model (rigid, flexible).

The research on the vehicle-road interaction in China is mainly carried in the universities. As early as 1965, Fang recognized that the vibration of the vehicle is caused by the pavement irregularity. Sun and Deng analyzed the influence of vehicle parameters on the dynamic load in detail by using the random theory based on a simplified quarter vehicle model. Lv established a heavy vehicle model for optimization of improved vehicle ride comfort and reduced tire dynamic shock to the ground. Lv' and Dong other achievements are summarized in the monograph entitled "Mechanical Analysis of Vehicle-Asphalt Pavement System". Lu investigated a heavy vehicle-tire-road coupled system and found that nonlinearities in the vehicle suspension contribute to the improved ride comfort and reduced dynamic tire force. Yang developed a 3D car-pavement-subgrade coupled model and studied the effect of vehicle-pavement coupling on the vehicle body acceleration, suspension deformation, tire force and the surface displacement by comparing the dynamical response of this model and the traditional vehicle-road model. A nonlinear vehicle-road coupled system involves a 7 DOF vehicle moving along a simply supported double-layer rectangular thin plate on a nonlinear viscoelastic foundation, figure It is found that the nonlinearity of vehicle and the viscoelasticity of road material should be considered when the vehicle-road system response is studied.

Despite of all these studies, the important issues about the interaction between the vehicle and road surface are still not resolved. The contact mechanism between the tire and road surface needs further investigations.

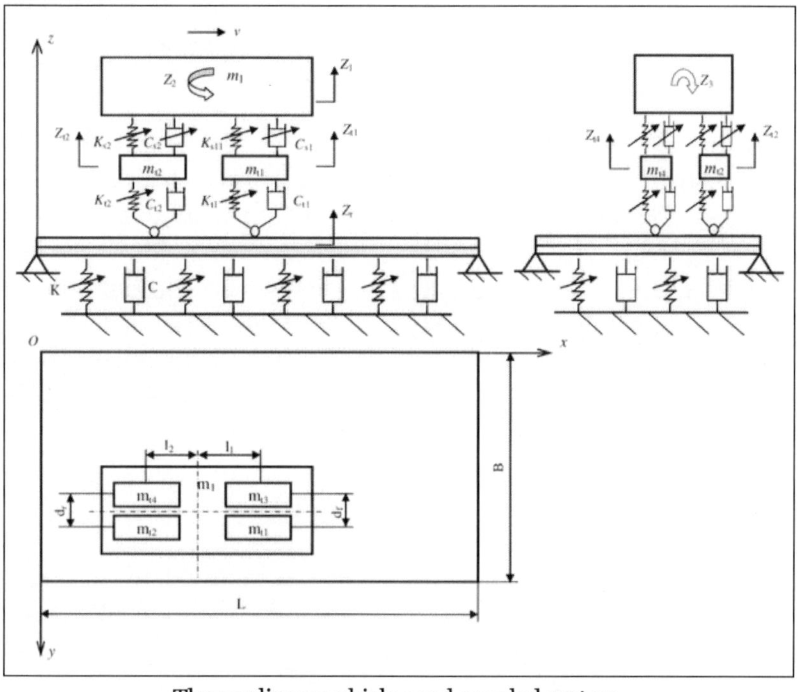

The nonlinear vehicle-road coupled system.

Kinetic Energy

The kinetic energy (KE) of an object is the energy that it possesses due to its motion. It is defined as the work needed to accelerate a body of a given mass from rest to its stated velocity. Having gained this energy during its acceleration, the body maintains this kinetic energy unless its speed changes. The same amount of work is done by the body when decelerating from its current speed to a state of rest.

In classical mechanics, the kinetic energy of a non-rotating object of mass m traveling at a speed v is $\frac{1}{2}mv^2$. In relativistic mechanics, this is a good approximation only when v is much less than the speed of light.

The standard unit of kinetic energy is the joule, while the imperial unit of kinetic energy is the foot-pound.

Energy occurs in many forms, including chemical energy, thermal energy, electromagnetic radiation, gravitational energy, electric energy, elastic energy, nuclear energy, and rest energy. These can be categorized in two main classes: potential energy and kinetic energy. Kinetic energy is the movement energy of an object. Kinetic energy can be transferred between objects and transformed into other kinds of energy.

Kinetic energy may be best understood by examples that demonstrate how it is transformed to and from other forms of energy. For example, a cyclist uses chemical energy provided by food to accelerate a bicycle to a chosen speed. On a level surface, this speed can be maintained without further work, except to overcome air resistance and friction. The chemical energy has been converted into kinetic energy, the energy of motion, but the process is not completely efficient and produces heat within the cyclist.

The kinetic energy in the moving cyclist and the bicycle can be converted to other forms. For example, the cyclist could encounter a hill just high enough to coast up, so that the bicycle comes to a complete halt at the top. The kinetic energy has now largely been converted to gravitational potential energy that can be released by freewheeling down the other side of the hill. Since the bicycle lost some of its energy to friction, it never regains all of its speed without additional pedaling. The energy is not destroyed; it has only been converted to another form by friction. Alternatively, the cyclist could connect a dynamo to one of the wheels and generate some electrical energy on the descent. The bicycle would be traveling slower at the bottom of the hill than without the generator because some of the energy has been diverted into electrical energy. Another possibility would be for the cyclist to apply the brakes, in which case the kinetic energy would be dissipated through friction as heat.

Like any physical quantity that is a function of velocity, the kinetic energy of an object depends on the relationship between the object and the observer's frame of reference. Thus, the kinetic energy of an object is not invariant.

Spacecraft use chemical energy to launch and gain considerable kinetic energy to reach orbital velocity. In an entirely circular orbit, this kinetic energy remains constant because there is almost no friction in near-earth space. However, it becomes apparent at re-entry when some of the kinetic energy is converted to heat. If the orbit is elliptical or hyperbolic, then throughout the orbit kinetic and potential energy are exchanged; kinetic energy is greatest and potential energy lowest at closest approach to the earth or other massive body, while potential energy is greatest and kinetic energy the lowest at maximum distance. Without loss or gain, however, the sum of the kinetic and potential energy remains constant.

Kinetic energy can be passed from one object to another. In the game of billiards, the player imposes kinetic energy on the cue ball by striking it with the cue stick. If the cue ball collides with another ball, it slows down dramatically, and the ball it hit accelerates its speed as the kinetic energy is passed on to it. Collisions in billiards are effectively elastic collisions, in which kinetic energy is preserved. In inelastic collisions, kinetic energy is dissipated in various forms of energy, such as heat, sound, binding energy (breaking bound structures).

Flywheels have been developed as a method of energy storage. This illustrates that kinetic energy is also stored in rotational motion.

Several mathematical descriptions of kinetic energy exist that describe it in the appropriate physical situation. For objects and processes in common human experience, the formula $\frac{1}{2}mv^2$ given by Newtonian (classical) mechanics is suitable. However, if the speed of the object is comparable to the speed of light, relativistic effects become significant and the relativistic formula is used. If the object is on the atomic or sub-atomic scale, quantum mechanical effects are significant, and a quantum mechanical model must be employed.

Newtonian Kinetic Energy

Kinetic Energy of Rigid Bodies

In classical mechanics, the kinetic energy of a point object (an object so small that its mass can be assumed to exist at one point), or a non-rotating rigid body depends on the mass of the body as well as its speed. The kinetic energy is equal to 1/2 the product of the mass and the square of the speed. In formula form:

$$E_{\text{k}} = \frac{1}{2}mv^2$$

Where m is the mass and v is the speed (or the velocity) of the body. In SI units, mass is measured in kilograms, speed in metres per second, and the resulting kinetic energy is in joules.

For example, one would calculate the kinetic energy of an 80 kg mass (about 180 lbs) traveling at 18 metres per second (about 40 mph, or 65 km/h) as:

$$E_{\text{k}} = \frac{1}{2} \cdot 80\,\text{kg} \cdot (18\,\text{m/s})^2 = 12{,}960\,\text{J} = 12.96\,\text{kJ}$$

When a person throws a ball, the person does work on it to give it speed as it leaves the hand. The moving ball can then hit something and push it, doing work on what it hits. The kinetic energy of a moving object is equal to the work required to bring it from rest to that speed, or the work the object can do while being brought to rest: net force × displacement = kinetic energy, i.e.,

$$Fs = \frac{1}{2}mv^2$$

Since the kinetic energy increases with the square of the speed, an object doubling its speed has four times as much kinetic energy. For example, a car traveling twice as fast as another requires four times as much distance to stop, assuming a constant braking force. As a consequence of this quadrupling, it takes four times the work to double the speed.

The kinetic energy of an object is related to its momentum by the equation:

$$E_{\text{k}} = \frac{p^2}{2m}$$

where:

- p is momentum.
- m is mass of the body.

For the translational kinetic energy, that is the kinetic energy associated with rectilinear motion, of a rigid body with constant mass m, whose center of mass is moving in a straight line with speed v, as seen above is equal to:

$$E_{\text{t}} = \frac{1}{2}mv^2$$

where:

- m is the mass of the body.
- v is the speed of the center of mass of the body.

The kinetic energy of any entity depends on the reference frame in which it is measured. However the total energy of an isolated system, i.e. one in which energy can neither enter nor leave, does not change over time in the reference frame in which it is measured. Thus, the chemical energy converted to kinetic energy by a rocket engine is divided differently between the rocket ship and its exhaust stream depending upon the chosen reference frame. This is called the Oberth effect. But the total energy of the system, including kinetic energy, fuel chemical energy, heat, etc., is conserved over time, regardless of the choice of reference frame. Different observers moving with different reference frames would however disagree on the value of this conserved energy.

The kinetic energy of such systems depends on the choice of reference frame: the reference frame that gives the minimum value of that energy is the center of momentum frame, i.e. the reference frame in which the total momentum of the system is zero. This minimum kinetic energy contributes to the invariant mass of the system as a whole.

Derivation

The work done in accelerating a particle with mass m during the infinitesimal time interval dt is given by the dot product of force F and the infinitesimal displacement dx.

$$\mathbf{F}\cdot d\mathbf{x} = \mathbf{F}\cdot \mathbf{v}dt = \frac{d\mathbf{p}}{dt}\cdot\mathbf{v}dt = \mathbf{v}\cdot d\mathbf{p} = \mathbf{v}\cdot d(m\mathbf{v}),$$

Where we have assumed the relationship p = m v and the validity of Newton's Second Law.

Applying the product rule we see that:

$$d(\mathbf{v}\cdot\mathbf{v}) = (d\mathbf{v})\cdot\mathbf{v} + \mathbf{v}\cdot(d\mathbf{v}) = 2(\mathbf{v}\cdot d\mathbf{v})$$

Therefore, (assuming constant mass so that dm = 0), we have,

$$\mathbf{v}\cdot d(m\mathbf{v}) = \frac{m}{2}d(\mathbf{v}\cdot\mathbf{v}) = \frac{m}{2}dv^2 = d\left(\frac{mv^2}{2}\right)$$

Since this is a total differential (that is, it only depends on the final state, not how the particle got there), we can integrate it and call the result kinetic energy. Assuming the object was at rest at time 0, we integrate from time 0 to time t because the work done by the force to bring the object from rest to velocity v is equal to the work necessary to do the reverse:

$$E_k = \int_0^t F\cdot d\mathbf{x} = \int_0^t v\cdot d(m\mathbf{v}) = \int_0^t d\left(\frac{mv^2}{2}\right) = \frac{mv^2}{2}.$$

This equation states that the kinetic energy (E_k) is equal to the integral of the dot product of the velocity (v) of a body and the infinitesimal change of the body's momentum (p). It is assumed that the body starts with no kinetic energy when it is at rest (motionless).

Rotating Bodies

If a rigid body Q is rotating about any line through the center of mass then it has rotational kinetic energy (E_r) which is simply the sum of the kinetic energies of its moving parts, and is thus given by:

$$E_r = \int_Q \frac{v^2 dm}{2} = \int_Q \frac{(r\omega)^2 dm}{2} = \frac{\omega^2}{2} \int_Q r^2 dm = \frac{\omega^2}{2} I = \frac{1}{2} I \omega^2$$

where:

- ω is the body's angular velocity.
- r is the distance of any mass dm from that line.
- I is the body's moment of inertia, equal to $\int_Q r^2 dm$.

(In this equation the moment of inertia must be taken about an axis through the center of mass and the rotation measured by ω must be around that axis; more general equations exist for systems where the object is subject to wobble due to its eccentric shape).

Kinetic Energy of Systems

A system of bodies may have internal kinetic energy due to the relative motion of the bodies in the system. For example, in the Solar System the planets and planetoids are orbiting the Sun. In a tank of gas, the molecules are moving in all directions. The kinetic energy of the system is the sum of the kinetic energies of the bodies it contains.

A macroscopic body that is stationary (i.e. a reference frame has been chosen to correspond to the body's center of momentum) may have various kinds of internal energy at the molecular or atomic level, which may be regarded as kinetic energy, due to molecular translation, rotation, and vibration, electron translation and spin, and nuclear spin. These all contribute to the body's mass, as provided by the special theory of relativity. When discussing movements of a macroscopic body, the kinetic energy referred to is usually that of the macroscopic movement only. However all internal energies of all types contribute to body's mass, inertia, and total energy.

Fluid Dynamics

In fluid dynamics, the kinetic energy per unit volume at each point in an incompressible fluid flow field is called the dynamic pressure at that point.

$$E_k = \frac{1}{2} m v^2$$

Dividing by V, the unit of volume:

$$\frac{E_\text{k}}{V} = \frac{1}{2}\frac{m}{V}v^2$$

$$q = \frac{1}{2}\rho v^2$$

where q is the dynamic pressure, and ρ is the density of the incompressible fluid.

Frame of Reference

The speed, and thus the kinetic energy of a single object is frame-dependent (relative): it can take any non-negative value, by choosing a suitable inertial frame of reference. For example, a bullet passing an observer has kinetic energy in the reference frame of this observer. The same bullet is stationary to an observer moving with the same velocity as the bullet, and so has zero kinetic energy. By contrast, the total kinetic energy of a system of objects cannot be reduced to zero by a suitable choice of the inertial reference frame, unless all the objects have the same velocity. In any other case, the total kinetic energy has a non-zero minimum, as no inertial reference frame can be chosen in which all the objects are stationary. This minimum kinetic energy contributes to the system's invariant mass, which is independent of the reference frame.

The total kinetic energy of a system depends on the inertial frame of reference: it is the sum of the total kinetic energy in a center of momentum frame and the kinetic energy the total mass would have if it were concentrated in the center of mass.

This may be simply shown: let $\mathbf{V}$ be the relative velocity of the center of mass frame i in the frame k. Since,

$$v^2 = (v_i + V)^2 = (\mathbf{v}_i + \mathbf{V})\cdot(\mathbf{v}_i + \mathbf{V}) = \mathbf{v}_i\cdot\mathbf{v}_i + 2\mathbf{v}_i\cdot\mathbf{V} + \mathbf{V}\cdot\mathbf{V} = v_i^2 + 2\mathbf{v}_i\cdot\mathbf{V} + V^2,$$

Then,

$$E_\text{k} = \int \frac{v^2}{2}dm = \int \frac{v_i^2}{2}dm + \mathbf{V}\cdot\int \mathbf{v}_i\, dm + \frac{V^2}{2}\int dm.$$

However, let $\int \frac{v_i^2}{2}dm = E_i$ the kinetic energy in the center of mass frame, $\int \mathbf{v}_i dm$ would be simply the total momentum that is by definition zero in the center of mass frame, and let the total mass: $\int dm = M$. Substituting, we get:

$$E_\text{k} = E_i + \frac{MV^2}{2}.$$

Thus the kinetic energy of a system is lowest to center of momentum reference frames, i.e., frames of reference in which the center of mass is stationary (either the center of mass frame or any other center of momentum frame). In any different frame of reference, there is additional kinetic energy corresponding to the total mass moving at the speed of the center of mass. The kinetic energy of the system in the center of momentum frame is a quantity that is invariant.

Rotation in Systems

It sometimes is convenient to split the total kinetic energy of a body into the sum of the body's center-of-mass translational kinetic energy and the energy of rotation around the center of mass (rotational energy):

$$E_\mathrm{k} = E_\mathrm{t} + E_\mathrm{r}$$

where:

- E_k is the total kinetic energy.
- E_t is the translational kinetic energy.
- E_r is the rotational energy or angular kinetic energy in the rest frame.

Thus the kinetic energy of a tennis ball in flight is the kinetic energy due to its rotation, plus the kinetic energy due to its translation.

Relativistic Kinetic Energy of Rigid Bodies

If a body's speed is a significant fraction of the speed of light, it is necessary to use relativistic mechanics to calculate its kinetic energy. In special relativity theory, the expression for linear momentum is modified.

With m being an object's rest mass, v and v its velocity and speed, and c the speed of light in vacuum, we use the expression for linear momentum $\mathbf{p} = m\gamma\mathbf{v}$, where:

$$\gamma = 1/\sqrt{1 - v^2/c^2}.$$

Integrating by parts yields:

$$E_\mathrm{k} = \int \mathbf{v} \cdot d\mathbf{p} = \int \mathbf{v} \cdot d(m\gamma\mathbf{v}) = m\gamma\mathbf{v} \cdot \mathbf{v} - \int m\gamma\mathbf{v} \cdot d\mathbf{v} = m\gamma v^2 - \frac{m}{2}\int \gamma d\left(v^2\right)$$

Since, $\gamma = \left(1 - v^2/c^2\right)^{-\frac{1}{2}}$

$$E_\mathrm{k} = m\gamma v^2 - \frac{-mc^2}{2}\int \gamma d\left(1 - \frac{v^2}{c^2}\right)$$

$$= m\gamma v^2 + mc^2\left(1 - \frac{v^2}{c^2}\right)^{\frac{1}{2}} - E_0$$

E_0 is a constant of integration for the indefinite integral.

Simplifying the expression we obtain,

$$\begin{aligned} E_{\mathrm{k}} &= m\gamma\left(v^2 + c^2\left(1 - \frac{v^2}{c^2}\right)\right) - E_0 \\ &= m\gamma\left(v^2 + c^2 - v^2\right) - E_0 \\ &= m\gamma c^2 - E_0 \end{aligned}$$

E_0 is found by observing that when $\mathbf{v} = 0$, $\gamma = 1$ and $E_{\mathrm{k}} = 0$, giving,

$$E_0 = mc^2$$

resulting in the formula,

$$E_{\mathrm{k}} = m\gamma c^2 - mc^2 = \frac{mc^2}{\sqrt{1 - \frac{v^2}{c^2}}} - mc^2$$

This formula shows that the work expended accelerating an object from rest approaches infinity as the velocity approaches the speed of light. Thus it is impossible to accelerate an object across this boundary.

The mathematical by-product of this calculation is the mass-energy equivalence formula—the body at rest must have energy content.

$$E_{\mathrm{rest}} = E_0 = mc^2$$

At a low speed ($v \ll c$), the relativistic kinetic energy is approximated well by the classical kinetic energy. This is done by binomial approximation or by taking the first two terms of the Taylor expansion for the reciprocal square root:

$$E_{\mathrm{k}} \approx mc^2\left(1 + \frac{1}{2}\frac{v^2}{c^2}\right) - mc^2 = \frac{1}{2}mv^2$$

So, the total energy E_k can be partitioned into the rest mass energy plus the Newtonian kinetic energy at low speeds.

When objects move at a speed much slower than light (e.g. in everyday phenomena on Earth), the first two terms of the series predominate. The next term in the Taylor series approximation:

$$E_{\mathrm{k}} \approx mc^2\left(1 + \frac{1}{2}\frac{v^2}{c^2} + \frac{3}{8}\frac{v^4}{c^4}\right) - mc^2 = \frac{1}{2}mv^2 + \frac{3}{8}m\frac{v^4}{c^2}$$

is small for low speeds. For example, for a speed of 10 km/s (22,000 mph) the correction to the Newtonian kinetic energy is 0.0417 J/kg (on a Newtonian kinetic energy of 50 MJ/kg) and for a speed of 100 km/s it is 417 J/kg (on a Newtonian kinetic energy of 5 GJ/kg).

The relativistic relation between kinetic energy and momentum is given by,

$$E_{\mathrm{k}} = \sqrt{p^2c^2 + m^2c^4} - mc^2$$

This can also be expanded as a Taylor series, the first term of which is the simple expression from Newtonian mechanics:

$$E_{\mathrm{k}} \approx \frac{p^2}{2m} - \frac{p^4}{8m^3c^2}.$$

This suggests that the formulae for energy and momentum are not special and axiomatic, but concepts emerging from the equivalence of mass and energy and the principles of relativity.

General Relativity

Using the convention that,

$$g_{\alpha\beta}\, u^{\alpha}\, u^{\beta} = -c^2$$

where the four-velocity of a particle is:

$$u^{\alpha} = \frac{dx^{\alpha}}{d\tau}$$

and τ is the proper time of the particle, there is also an expression for the kinetic energy of the particle in general relativity.

If the particle has momentum,

$$p_{\beta} = m\, g_{\beta\alpha}\, u^{\alpha}$$

as it passes by an observer with four-velocity u_{obs}, then the expression for total energy of the particle as observed (measured in a local inertial frame) is:

$$E = -\, p_{\beta}\, u^{\beta}_{\mathrm{obs}}$$

and the kinetic energy can be expressed as the total energy minus the rest energy:

$$E_k = -\, p_{\beta}\, u^{\beta}_{\mathrm{obs}} - m\, c^2 .$$

Consider the case of a metric that is diagonal and spatially isotropic (gtt, gss, gss, gss). Since,

$$u^{\alpha} = \frac{dx^{\alpha}}{dt}\frac{dt}{d\tau} = v^{\alpha}u^{t}$$

where v^{α} is the ordinary velocity measured w.r.t. the coordinate system, we get:

$$-c^2 = g_{\alpha\beta}u^{\alpha}u^{\beta} = g_{tt}\left(u^{t}\right)^2 + g_{ss}v^2\left(u^{t}\right)^2 .$$

Solving for u^t gives,

$$u^t = c\sqrt{\frac{-1}{g_{tt} + g_{ss}v^2}}.$$

Thus for a stationary observer (v = 0),

$$u^t_{\text{obs}} = c\sqrt{\frac{-1}{g_{tt}}}$$

and thus the kinetic energy takes the form,

$$E_{\text{k}} = -mg_{tt}u^t u^t_{\text{obs}} - mc^2 = mc^2\sqrt{\frac{g_{tt}}{g_{tt} + g_{ss}v^2}} - mc^2.$$

Factoring out the rest energy gives:

$$E_{\text{k}} = mc^2\left(\sqrt{\frac{g_{tt}}{g_{tt} + g_{ss}v^2}} - 1\right).$$

This expression reduces to the special relativistic case for the flat-space metric where,

$$g_{tt} = -c^2$$
$$g_{ss} = 1.$$

In the Newtonian approximation to general relativity,

$$g_{tt} = -\left(c^2 + 2\Phi\right)$$
$$g_{ss} = 1 - \frac{2\Phi}{c^2}$$

where Φ is the Newtonian gravitational potential. This means clocks run slower and measuring rods are shorter near massive bodies.

Kinetic Energy in Quantum Mechanics

In quantum mechanics, observables like kinetic energy are represented as operators. For one particle of mass m, the kinetic energy operator appears as a term in the Hamiltonian and is defined in terms of the more fundamental momentum operator $\hat{p}$. The kinetic energy operator in the non-relativistic case can be written as:

$$\hat{T} = \frac{\hat{p}^2}{2m}.$$

Notice that this can be obtained by replacing p by $\hat{p}$ in the classical expression for kinetic energy in terms of momentum,

$$E_k = \frac{p^2}{2m}.$$

In the Schrödinger picture, $\hat{p}$ takes the form $-i\hbar\nabla$ where the derivative is taken with respect to position coordinates and hence,

$$\hat{T} = -\frac{\hbar^2}{2m}\nabla^2.$$

The expectation value of the electron kinetic energy, $\langle \hat{T} \rangle$ for a system of N electrons described by the wavefunction $|\psi\rangle$is a sum of 1-electron operator expectation values:

$$\langle \hat{T} \rangle = \left\langle \psi \left| \sum_{i=1}^{N} \frac{-\hbar^2}{2m_e} \nabla_i^2 \right| \psi \right\rangle = -\frac{\hbar^2}{2m_e} \sum_{i=1}^{N} \langle \psi | \nabla_i^2 | \psi \rangle$$

where m_e is the mass of the electron and ∇_i^2 is the Laplacian operator acting upon the coordinates of the ith electron and the summation runs over all electrons.

The density functional formalism of quantum mechanics requires knowledge of the electron density only, i.e., it formally does not require knowledge of the wavefunction. Given an electron density $\rho(\mathbf{r})$, the exact N-electron kinetic energy functional is unknown; however, for the specific case of a 1-electron system, the kinetic energy can be written as,

$$T[\rho] = \frac{1}{8} \int \frac{\nabla\rho(\mathbf{r}) \cdot \nabla\rho(\mathbf{r})}{\rho(\mathbf{r})} d^3r$$

where $T[\rho]$ is known as the von Weizsäcker kinetic energy functional.

Free Body Diagram

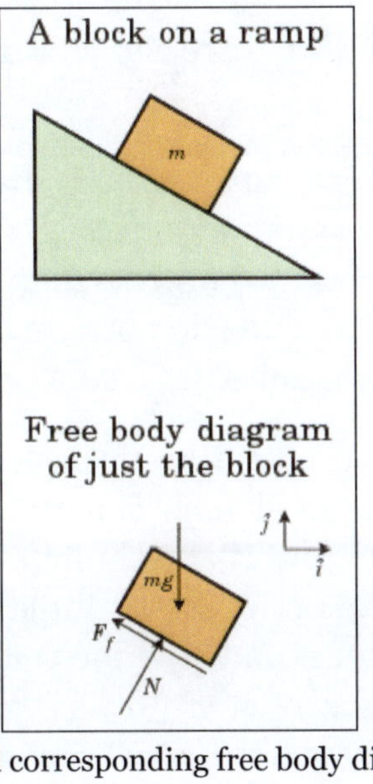

Block on a ramp and corresponding free body diagram of the block.

A free body diagram (force diagram, or FBD) is a graphical illustration used to visualize the applied forces, moments, and resulting reactions on a body in a given condition. They depict a body or connected bodies with all the applied forces and moments, and reactions, which act on the body(ies). The body may consist of multiple internal members (such as a truss), or be a compact body (such as a beam). A series of free bodies and other diagrams may be necessary to solve complex problems.

Purpose

Free body diagrams are used to visualize the forces and moments applied to a body and calculate the resulting reactions, in many types of mechanics problems. Most free body diagrams are used both to determine the loading of individual structural components and to calculate internal forces within the structure in almost all engineering disciplines from Biomechanics to Structural body. In the educational environment, learning to draw a free body diagram is an important step in understanding certain topics in physics, such as statics, dynamics and other forms of classical mechanics.

Features

A free body diagram is not meant to be a scaled drawing. It is a diagram that is modified as the problem is solved. There is an art and flexibility to the process. The iconography of a free body diagram, not only how it is drawn but also how it is interpreted, depends upon how a body is modeled.

Free body diagrams consist of:

- A simplified version of the body (often a dot or a box).
- Forces shown as straight arrows pointing in the direction they act on the body.
- Moments shown as curved arrows pointing in the direction they act on the body.
- A coordinate system.
- Frequently reaction to applied forces are shown with hash marks through the stem of the arrow.

The number of forces and moments shown in a free body diagram depends on the specific problem and the assumptions made; common assumptions are neglecting air resistance, friction and assuming rigid bodies. In statics all forces and moments must balance to zero; the physical interpretation of this is that if the forces and moments do not sum to zero the body is accelerating and the principles of statics do not apply. In dynamics the resultant forces and moments can be non-zero.

Free body diagrams may not represent an entire physical body. Using what is known as a "cut" only portions of a body are selected for modeling. This technique exposes internal forces, making them external, therefore allowing analysis. This technique is often used several times, iteratively to peel back forces acting on a physical body. For example, a gymnast performing the iron cross: analyzing the ropes and the person lets you know the total force (body weight, neglecting rope weight, breezes, buoyancy, electrostatics, relativity, rotation of the earth, etc.). Then cut the person out and only show one rope. You get force direction. Then only look at the person, now you can get hand forces. Now only look at the arm to get the shoulder forces and moments, and on and on until the component you intend to analyze is exposed.

Modeling the Body

A body may be modeled in three ways:

- A particle: This model may be used when any rotational effects are zero or have no interest even though the body itself may be extended. The body may be represented by a small symbolic blob and the diagram reduces to a set of concurrent arrows. A force on a particle is a bound vector.
- Rigid extended: Stresses and strains are of no interest but turning effects are. A force arrow should lie along the line of force, but where along the line is irrelevant. A force on an extended rigid body is a sliding vector.
- Non-rigid extended: The point of application of a force becomes crucial and has to be indicated on the diagram. A force on a non-rigid body is a bound vector. Some use the tail of the arrow to indicate the point of application. Others use the tip.

Example: A Body in Free Fall.

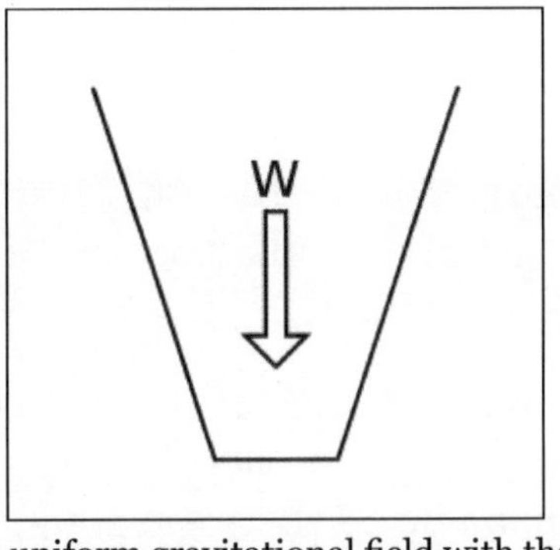

An empty rigid bucket in free fall in a uniform gravitational field with the force arrow at the center of gravity.

Consider a body in free fall in a uniform gravitational field. The body may be:

- A particle: It is enough to show a single vertically downward pointing arrow attached to a blob.
- Rigid extended: A single arrow suffices to represent the weight W even though gravitational attraction acts on every particle of the body.
- Non-rigid extended: In non-rigid analysis, it would be an error to associate a single point of application with the gravitational force.

What is included:

An FBD represents the body of interest and the external forces on it.

- The body: This is usually sketched in a schematic way depending on the body—particle/extended, rigid/non-rigid—and on what questions are to be answered. Thus if rotation of the body and torque is in consideration, an indication of size and shape of the body is needed. For example, the brake dive of a motorcycle cannot be found from a single point, and a sketch with finite dimensions is required.

- The external forces: These are indicated by labelled arrows. In a fully solved problem, a force arrow is capable of indicating:
  - The direction and the line of action.
  - The magnitude.
  - The point of application.
  - A reaction as opposed to an applied load if a hash is present through the arrow.

Typically, however, a provisional free body sketch is drawn before all these things are known. After all, the whole point of the diagram is to help to determine magnitude, direction, and point of application of the external loads. Thus when a force arrow is originally drawn its length may not be meant to indicate the unknown magnitude. Its line may not correspond to the exact line of action. Even its direction may turn out to be wrong. Very often the original direction of the arrow may be directly opposite to the true direction. External forces known to be small that are known to have negligible effect on the result of the analysis are sometimes omitted, but only after careful consideration or after other analysis proving it (e.g. buoyancy forces of the air in the analysis of a chair, or atmospheric pressure on the analysis of a frying pan).

The external forces acting on the object include friction, gravity, normal force, drag, tension, or a human force due to pushing or pulling. When in a non-inertial reference frame, fictitious forces, such as centrifugal pseudoforce are appropriate.

A coordinate system is sometimes included, and is chosen according to convenience (or advantage). Savvy selection of coordinate frame may make defining the vectors simpler when writing the equations of motion. The x direction might be chosen to point down the ramp in an inclined plane problem, for example. In that case the friction force only has an x component, and the normal force only has a y component. The force of gravity will still have components in both the x and y direction: $mg\sin(\theta)$ in the x and $mg\cos(\theta)$ in the y, where θ is the angle between the ramp and the horizontal.

Exclusions

There are some things that a free body diagram explicitly excludes. Although other sketches that include these things may be helpful in visualizing a problem, a proper free body diagram should *not* show:

- Bodies other than the free body.
- Constraints:
  - The body is not free from constraints; the constraints have just been replaced by the forces and moments that they exert on the body.

- Forces exerted *by* the free body:
  - A diagram showing the forces exerted both on *and* by a body is likely to be confusing since all the forces will cancel out. By Newton's 3rd law if body *A* exerts a force on body *B* then *B* exerts an equal and opposite force on *A*. This should not be confused with the equal and opposite forces that are necessary to hold a body in equilibrium.
- Internal forces:
  - For example, if an entire truss is being analyzed, the forces between the individual truss members are not included.
- Velocity or acceleration vectors.

Analysis

A free body diagram is analyzed by summing all of the forces, often accomplished by summing the forces in each of the axis directions. When the net force is zero, the body must be at rest or must be moving at a constant velocity (constant speed and direction), by Newton's first law. If the net force is not zero, then the body is accelerating in that direction according to Newton's second law.

Angled Forces

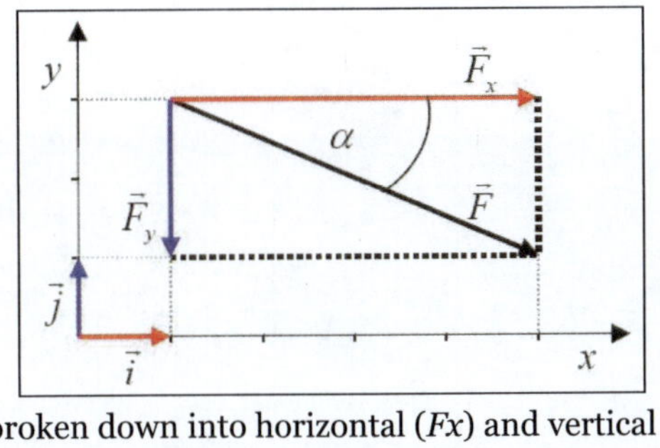

Angled force (*F*) broken down into horizontal (*Fx*) and vertical (*Fy*) components.

Determining the sum of the forces is straightforward if all they are aligned with the coordinate frame's axes, but it is somewhat more complex if some forces are not aligned. It is often convenient to analyze the components of the forces, in which case the symbols ΣF_x and ΣF_y are used instead of ΣF. Forces that point at an angle to the diagram's coordinate axis can be broken down into two parts (or three, for three dimensional problems)—each part being directed along one of the axes—horizontally (*Fx*) and vertically (*Fy*).

Example: A Block on an Inclined Plane

A simple free body diagram, shown above, of a block on a ramp illustrates this.

- All external supports and structures have been replaced by the forces they generate. These include:
  - *mg*: the product of the mass of the block and the constant of gravitation acceleration: its weight.
  - *N*: the normal force of the ramp.

- *Ff*: the friction force of the ramp.

- The force vectors show direction and point of application and are labeled with their magnitude.
- It contains a coordinate system that can be used when describing the vectors.

Some care is needed in interpreting the diagram:

- The normal force has been shown to act at the midpoint of the base, but if the block is in static equilibrium its true location is directly below the center of mass, where the weight acts, because that is necessary to compensate for the moment of the friction.
- Unlike the weight and normal force, which are expected to act at the tip of the arrow, the friction force is a sliding vector and thus the point of application is not relevant, and the friction acts along the whole base.

Kinetic Diagram

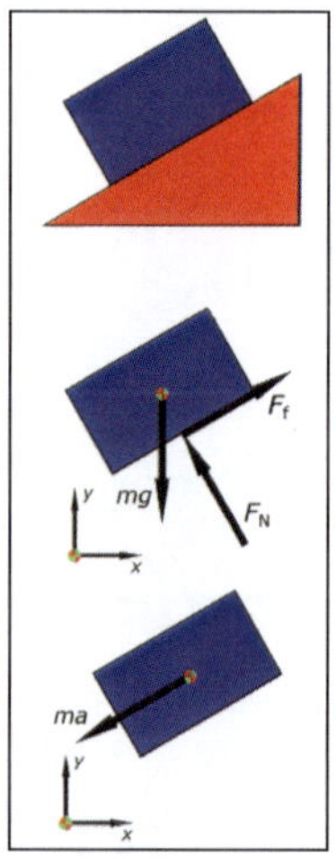

Free body and kinetic diagrams of an inclined block.

In dynamics a kinetic diagram is a pictorial device used in analyzing mechanics problems when there is determined to be a net force and moment acting on a body. They are related to and often used with free body diagrams, but depict only the net force and moment rather than all of the forces being considered.

Kinetic diagrams are not required to solve dynamics problems; their use in teaching dynamics is argued against by some in favor of other methods that they view as simpler. They appear in some dynamics texts but are absent in others.

Kinematics

Kinematics is a subdivision of classical mechanics concerned with the geometrically possible motion of a body or system of bodies without consideration of the forces involved (i.e., causes and effects of the motions).

Kinematics aims to provide a description of the spatial position of bodies or systems of material particles, the rate at which the particles are moving (velocity), and the rate at which their velocity is changing (acceleration). When the causative forces are disregarded, motion descriptions are possible only for particles having constrained motion—i.e., moving on determinate paths. In unconstrained, or free, motion, the forces determine the shape of the path.

For a particle moving on a straight path, a list of positions and corresponding times would constitute a suitable scheme for describing the motion of the particle. A continuous description would require a mathematical formula expressing position in terms of time.

When a particle moves on a curved path, a description of its position becomes more complicated and requires two or three dimensions. In such cases continuous descriptions in the form of a single graph or mathematical formula are not feasible. The position of a particle moving on a circle, for example, can be described by a rotating radius of the circle, like the spoke of a wheel with one end fixed at the centre of the circle and the other end attached to the particle. The rotating radius is known as a position vector for the particle, and, if the angle between it and a fixed radius is known as a function of time, the magnitude of the velocity and acceleration of the particle can be calculated. Velocity and acceleration, however, have direction as well as magnitude; velocity is always tangent to the path, while acceleration has two components, one tangent to the path and the other perpendicular to the tangent.

A formal study of physics begins with kinematics. Kinematic analysis is the process of measuring the kinematic quantities used to describe motion. The study of kinematics can be abstracted into purely mathematical expressions, which can be used to calculate various aspects of motion such as velocity, acceleration, displacement, time, and trajectory.

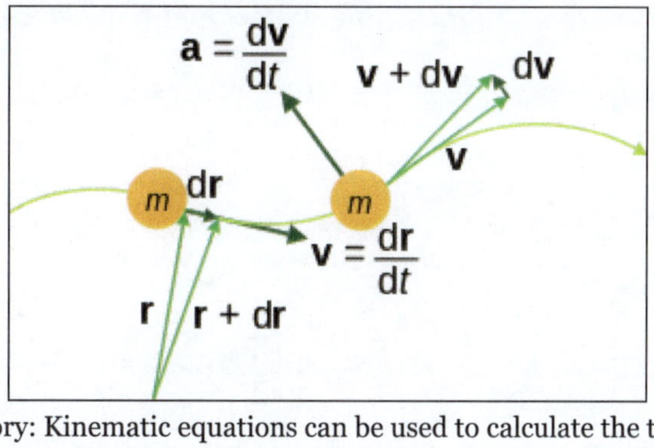

Kinematics of a particle trajectory: Kinematic equations can be used to calculate the trajectory of particles or objects. The physical quantities relevant to the motion of a particle include: Mass m, position r, velocity v, acceleration a.

Reference Frames and Displacement

In order to describe an object's motion, you need to specify its position relative to a convenient reference frame.

In order to describe the motion of an object, you must first describe its position – where it is at any particular time. More precisely, you need to specify its position relative to a convenient reference frame. Earth is often used as a reference frame, and we often describe the position of objects related to its position to or from Earth. Mathematically, the position of an object is generally represented by the variable x.

Frames of Reference

There are two choices you have to make in order to define a position variable x. You have to decide where to put x = 0 and which direction will be positive. This is referred to as choosing a coordinate system, or choosing a frame of reference. As long as you are consistent, any frame is equally valid. But you don't want to change coordinate systems in the middle of a calculation. Imagine sitting in a train in a station when suddenly you notice that the station is moving backward. Most people would say that they just failed to notice that the train was moving — it only seemed like the station was moving. But this shows that there is a third arbitrary choice that goes into choosing a coordinate system: valid frames of reference can differ from each other by moving relative to one another. It might seem strange to use a coordinate system moving relative to the earth — but, for instance, the frame of reference moving along with a train might be far more convenient for describing things happening inside the train. Frames of reference are particularly important when describing an object's displacement.

Displacement

Displacement is the change in position of an object relative to its reference frame. For example, if a car moves from a house to a grocery store, its displacement is the relative distance of the grocery store to the reference frame, or the house. The word "displacement" implies that an object has moved or has been displaced. Displacement is the change in position of an object and can be represented mathematically as follows:

$$\Delta x = x_f - x_0$$

where Δx is displacement, xf is the final position, and x_0 is the initial position.

Shows the importance of using a frame of reference when describing the displacement of a passenger on an airplane.

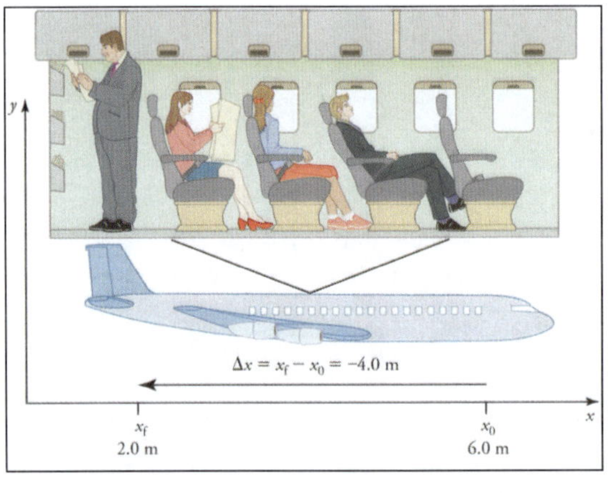

Displacement in Terms of Frame of Reference: A passenger moves from his seat to the back of the plane. His location relative to the airplane is given by x. The -4.0m displacement of the passenger relative to the plane is represented by an arrow toward the rear of the plane. Notice that the arrow representing his displacement is twice as long as the arrow representing the displacement of the professor (he moves twice as far).

References

- Dynamics-physics, science: britannica.com, Retrieved 27 February, 2019
- Dynamics: physics.info, Retrieved 16 June, 2019
- Klerk, D. De; Rixen, D. J.; Voormeeren, S. N. (2008-01-01). "General Framework for Dynamic Substructuring: History, Review and Classification of Techniques". AIAA Journal. 46 (5): 1169–1181. Bibcode:2008AIAAJ..46.1169D. doi:10.2514/1.33274. ISSN 0001-1452
- Introduction-to-rotational-dynamics, system-of-particles-and-rotational-dynamics: toppr.com, Retrieved 25 January, 2019
- John Robert Taylor (2005). Classical Mechanics. University Science Books. ISBN 978-1-891389-22-1
- Kinematics, science: britannica.com, Retrieved 16 May, 2019
- Chen, W. J., Gunter, E. J. (2005). Introduction to Dynamics of Rotor-Bearing Systems. Victoria, BC: Trafford. ISBN 978-1-4120-5190-3
- Basics-of-kinematics, boundless-physics: courses.lumenlearning.com, Retrieved 28 April, 2019

Rigid Body Mechanics 5

- **Rigid Body**
- **Rigid Body Mechanics**
- **Rotational Motion of a Rigid Body**
- **Rigid Body Dynamics**
- **Rigid Line Inclusion**
- **Rigid Rotor**
- **Euler's Equations**

A rigid body refers to a solid body in which deformation is zero or negligible. Rigid body mechanics is the study of the movement of interconnected bodies influenced by the external forces. This chapter has been carefully written to provide an extensive understanding of the subject.

Rigid Body

A rigid body is an extended area of material that includes all the points inside it, and which moves so that the distances and angles between all its points remain constant. The location of a rigid body can be described by the position of one point P inside it, together with the rotation angle of the body (one angle in 2D, three angles in 3D).

	Location Description	Velocity Description
Point Mass	position vector $\vec{r}p$	velocity vector $\vec{v}p$
Rigid Body In 2D	position vector $\vec{r}p$ angle θ	velocity vector $\vec{v}p$ angular velocity ω
Rigid Body In 3D	position vector $\vec{r}p$ angles θ, ϕ, ψ	velocity vector $\vec{v}p$ angular velocity vector $\vec{\omega}$

Neither point masses nor rigid bodies can physically exist, as no body can really be a single point with no extent, and no extended body can be exactly rigid. Despite this, these are very useful models for mechanics and dynamics.

Rotation Angles for Rigid Bodies

All points on a rigid body have the same angular rotation angles, Because the angular velocity is the derivative of the rotation angles, this means that every point on a rigid body has the same angular velocity $\vec{\omega}$, and also the same angular acceleration $\vec{\alpha}$.

In 2D the angle θ of a rigid body the angle of rotation from a fixed reference (typically the $\hat{i}$ direction), measured positive counter-clockwise. The angular velocity is $\omega = \dot{\theta}$ and the angular acceleration is $\alpha = \dot{\omega} = \ddot{\theta}$. The vector versions of these are $\omega = \omega\hat{k}$ and $\vec{\alpha} = \alpha\hat{k}$, where $\hat{k}$ is the out-of-plane direction.

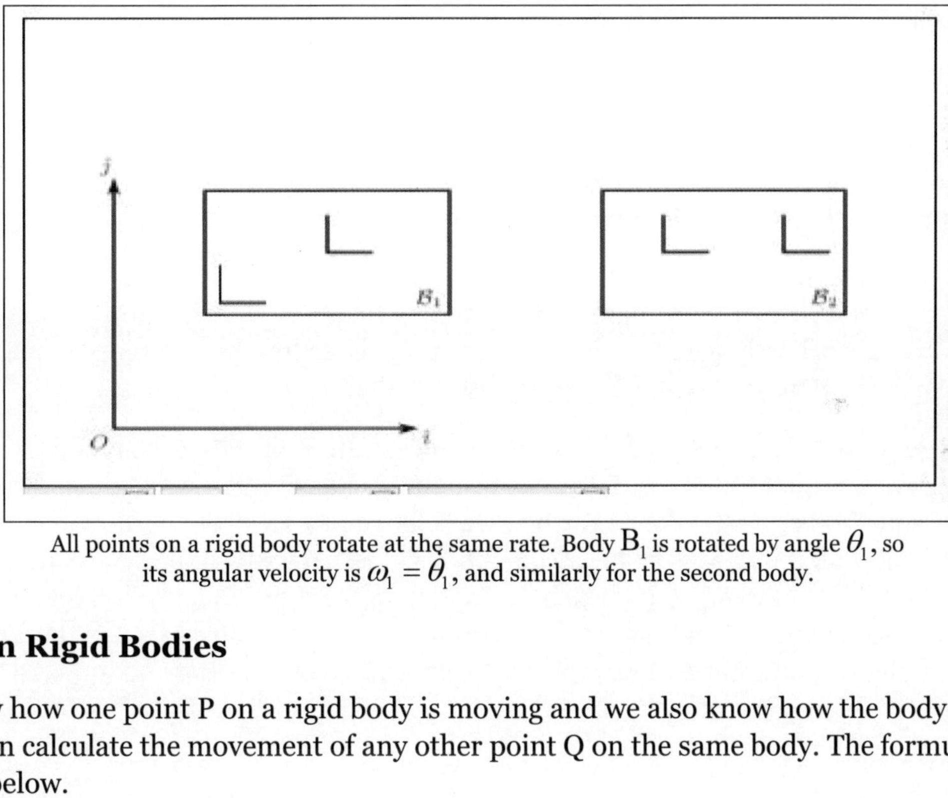

All points on a rigid body rotate at the same rate. Body B_1 is rotated by angle θ_1, so its angular velocity is $\omega_1 = \theta_1$, and similarly for the second body.

Points on Rigid Bodies

If we know how one point P on a rigid body is moving and we also know how the body is rotating, then we can calculate the movement of any other point Q on the same body. The formulas for this are given below.

Rigid Body Point Relations

$$\vec{r}_Q = \vec{r}_P + \vec{r}_{PQ}$$

$$\vec{v}_Q = \vec{v}_P + \vec{\omega} \times \vec{r}_{PQ}$$

$$\vec{\alpha}_Q = \vec{\alpha}_P + \vec{\alpha} \times \vec{r}_{PQ} + \vec{\omega} \times \left(\vec{\omega} \times \vec{r}_{PQ}\right)$$

Points P and Q are two locations on a rigid body. Vectors $\vec{\omega}$ and $\vec{\alpha}$ are the angular velocity and angular acceleration of the rigid body.

Kinematics

Linear and Angular Position

The position of a rigid body is the position of all the particles of which it is composed. To simplify the description of this position, we exploit the property that the body is rigid, namely that all its particles maintain the same distance relative to each other. If the body is rigid, it is sufficient to describe the position of at least three non-collinear particles. This makes it possible to reconstruct the position of all the other particles, provided that their time-invariant position relative to the three selected particles is known. However, typically a different, mathematically more convenient, but equivalent approach is used. The position of the whole body is represented by:

- The linear position or position of the body, namely the position of one of the particles of the body, specifically chosen as a reference point (typically coinciding with the center of mass or centroid of the body), together with.
- The angular position (also known as orientation, or attitude) of the body.

Thus, the position of a rigid body has two components: Linear and angular, respectively. The same is true for other kinematic and kinetic quantities describing the motion of a rigid body, such as linear and angular velocity, acceleration, momentum, impulse, and kinetic energy.

The linear position can be represented by a vector with its tail at an arbitrary reference point in space (the origin of a chosen coordinate system) and its tip at an arbitrary point of interest on the rigid body, typically coinciding with its center of mass or centroid. This reference point may define the origin of a coordinate system fixed to the body.

There are several ways to numerically describe the orientation of a rigid body, including a set of three Euler angles, a quaternion, or a direction cosine matrix (also referred to as a rotation matrix). All these methods actually define the orientation of a basis set (or coordinate system) which has a fixed orientation relative to the body (i.e. rotates together with the body), relative to another basis set (or coordinate system), from which the motion of the rigid body is observed. For instance, a basis set with fixed orientation relative to an airplane can be defined as a set of three orthogonal unit vectors b_1, b_2, b_3, such that b1 is parallel to the chord line of the wing and directed forward, b_2 is normal to the plane of symmetry and directed rightward, and b_3 is given by the cross product $b_3 = b_1 \times b_2$.

In general, when a rigid body moves, both its position and orientation vary with time. In the kinematic sense, these changes are referred to as translation and rotation, respectively. Indeed, the position of a rigid body can be viewed as a hypothetic translation and rotation (roto-translation) of the body starting from a hypothetic reference position (not necessarily coinciding with a position actually taken by the body during its motion).

Linear and Angular Velocity

Velocity (also called linear velocity) and angular velocity are measured with respect to a frame of reference.

The linear velocity of a rigid body is a vector quantity, equal to the time rate of change of its linear position. Thus, it is the velocity of a reference point fixed to the body. During purely translational motion (motion with no rotation), all points on a rigid body move with the same velocity. However, when motion involves rotation, the instantaneous velocity of any two points on the body will generally not be the same. Two points of a rotating body will have the same instantaneous velocity only if they happen to lie on an axis parallel to the instantaneous axis of rotation.

Angular velocity is a vector quantity that describes the angular speed at which the orientation of the rigid body is changing and the instantaneous axis about which it is rotating (the existence of this instantaneous axis is guaranteed by the Euler's rotation theorem). All points on a rigid body experience the same angular velocity at all times. During purely rotational motion, all points on the body change position except for those lying on the instantaneous axis of rotation. The relationship between orientation and angular velocity is not directly analogous to the relationship between position and velocity. Angular velocity is not the time rate of change of orientation, because there is no such concept as an orientation vector that can be differentiated to obtain the angular velocity.

Kinematical Equations

Addition Theorem for Angular Velocity

The angular velocity of a rigid body B in a reference frame N is equal to the sum of the angular velocity of a rigid body D in N and the angular velocity of B with respect to D:

$$^{N}\omega^{B} = {}^{N}\omega^{D} + {}^{D}\omega^{B}.$$

In this case, rigid bodies and reference frames are indistinguishable and completely interchangeable.

Addition Theorem for Position

For any set of three points P, Q, and R, the position vector from P to R is the sum of the position vector from P to Q and the position vector from Q to R:

$$\mathbf{r}^{PR} = \mathbf{r}^{PQ} + \mathbf{r}^{QR}.$$

Mathematical Definition of Velocity

The velocity of point P in reference frame N is defined as the time derivative in N of the position vector from O to P:

$$^{N}\mathbf{v}^{P} = \frac{^{N}\mathrm{d}}{\mathrm{d}t}(\mathbf{r}^{OP})$$

where O is any arbitrary point fixed in reference frame N, and the N to the left of the d/dt operator indicates that the derivative is taken in reference frame N. The result is independent of the selection of O so long as O is fixed in N.

Mathematical Definition of Acceleration

The acceleration of point P in reference frame N is defined as the time derivative in N of its velocity:

$$ {}^{N}\mathbf{a}^{P} = \frac{{}^{N}d}{dt}({}^{N}\mathbf{v}^{P}). $$

Velocity of Two Points Fixed on a Rigid Body

For two points P and Q that are fixed on a rigid body B, where B has an angular velocity ${}^{N}\omega^{B}$ in the reference frame N, the velocity of Q in N can be expressed as a function of the velocity of P in N:

$$ {}^{N}\mathbf{v}^{Q} = {}^{N}\mathbf{v}^{P} + {}^{N}\omega^{B} \times \mathbf{r}^{PQ}. $$

Acceleration of Two Points Fixed on a Rigid Body

By differentiating the equation for the Velocity of two points fixed on a rigid body in N with respect to time, the acceleration in reference frame N of a point Q fixed on a rigid body B can be expressed as:

$$ {}^{N}\mathbf{a}^{Q} = {}^{N}\mathbf{a}^{P} + {}^{N}\omega^{B} \times \left({}^{N}\omega^{B} \times \mathbf{r}^{PQ}\right) + {}^{N}\alpha^{B} \times \mathbf{r}^{PQ} $$

Where ${}^{N}\alpha^{B}$ is the angular acceleration of B in the reference frame N.

Angular Velocity and Acceleration of two Points Fixed on a Rigid Body

All points on a rigid body B have the same angular velocity ${}^{N}\omega^{B}$ in a fixed reference frame N, and thus the same angular acceleration ${}^{N}\alpha^{B}$.

Velocity of one Point Moving on a Rigid Body

If the point R is moving in rigid body B while B moves in reference frame N, then the velocity of R in N is,

$$ {}^{N}\mathbf{v}^{R} = {}^{N}\mathbf{v}^{Q} + {}^{B}\mathbf{v}^{R} $$

Where Q is the point fixed in B that is instantaneously coincident with R at the instant of interest. This relation is often combined with the relation for the Velocity of two points fixed on a rigid body.

Acceleration of one Point Moving on a Rigid Body

The acceleration in reference frame N of the point R moving in body B while B is moving in frame N is given by,

$$ {}^{N}\mathbf{a}^{R} = {}^{N}\mathbf{a}^{Q} + {}^{B}\mathbf{a}^{R} + 2\,{}^{N}\omega^{B} \times {}^{B}\mathbf{v}^{R} $$

Where Q is the point fixed in B that instantaneously coincident with R at the instant of interest. This equation is often combined with Acceleration of two points fixed on a rigid body.

Other Quantities

If C is the origin of a local coordinate system L, attached to the body,

- The spatial or twist acceleration of a rigid body is defined as the spatial acceleration of C (as opposed to material acceleration above);

$$\psi(t,\mathbf{r}_0) = \mathbf{a}(t,\mathbf{r}_0) - \omega(t)\times\mathbf{v}(t,\mathbf{r}_0) = \psi_c(t) + \alpha(t)\times A(t)\mathbf{r}_0$$

Where,

- $\mathbf{r}_0$ represents the position of the point/particle with respect to the reference point of the body in terms of the local coordinate system L (the rigidity of the body means that this does not depend on time).
- $A(t)$ is the orientation matrix, an orthogonal matrix with determinant 1, representing the orientation (angular position) of the local coordinate system L, with respect to the arbitrary reference orientation of another coordinate system G. Think of this matrix as three orthogonal unit vectors, one in each column, which define the orientation of the axes of L with respect to G.
- $\omega(t)$ represents the angular velocity of the rigid body.
- $\mathbf{v}(t,\mathbf{r}_0)$ represents the total velocity of the point/particle.
- $\mathbf{a}(t,\mathbf{r}_0)$ represents the total acceleration of the point/particle.
- $\alpha(t)$ represents the angular acceleration of the rigid body.
- $\psi(t,\mathbf{r}_0)$ represents the spatial acceleration of the point/particle.
- $\psi_c(t)$ represents the spatial acceleration of the rigid body (i.e. the spatial acceleration of the origin of L).

In 2D, the angular velocity is a scalar, and matrix A(t) simply represents a rotation in the xy-plane by an angle which is the integral of the angular velocity over time.

Vehicles, walking people, etc., usually rotate according to changes in the direction of the velocity: they move forward with respect to their own orientation. Then, if the body follows a closed orbit in a plane, the angular velocity integrated over a time interval in which the orbit is completed once, is an integer times 360°. This integer is the winding number with respect to the origin of the velocity. Compare the amount of rotation associated with the vertices of a polygon.

Kinetics

Any point that is rigidly connected to the body can be used as reference point (origin of coordinate system L) to describe the linear motion of the body (the linear position, velocity and acceleration vectors depend on the choice).

However, depending on the application, a convenient choice may be:

- The center of mass of the whole system, which generally has the simplest motion for a body moving freely in space;

- A point such that the translational motion is zero or simplified, e.g. on an axle or hinge, at the center of a ball and socket joint, etc.

When the center of mass is used as reference point:

- The (linear) momentum is independent of the rotational motion. At any time it is equal to the total mass of the rigid body times the translational velocity.
- The angular momentum with respect to the center of mass is the same as without translation: At any time it is equal to the inertia tensor times the angular velocity. When the angular velocity is expressed with respect to a coordinate system coinciding with the principal axes of the body, each component of the angular momentum is a product of a moment of inertia (a principal value of the inertia tensor) times the corresponding component of the angular velocity; the torque is the inertia tensor times the angular acceleration.
- Possible motions in the absence of external forces are translation with constant velocity, steady rotation about a fixed principal axis, and also torque-free precession.
- The net external force on the rigid body is always equal to the total mass times the translational acceleration (i.e., Newton's second law holds for the translational motion, even when the net external torque is nonzero, and the body rotates).
- The total kinetic energy is simply the sum of translational and rotational energy.

Geometry

Two rigid bodies are said to be different (not copies) if there is no proper rotation from one to the other. A rigid body is called chiral if its mirror image is different in that sense, i.e., if it has either no symmetry or its symmetry group contains only proper rotations. In the opposite case an object is called achiral: the mirror image is a copy, not a different object. Such an object may have a symmetry plane, but not necessarily: there may also be a plane of reflection with respect to which the image of the object is a rotated version. The latter applies for *S2n*, of which the case $n = 1$ is inversion symmetry.

For a (rigid) rectangular transparent sheet, inversion symmetry corresponds to having on one side an image without rotational symmetry and on the other side an image such that what shines through is the image at the top side, upside down. We can distinguish two cases:

- The sheet surface with the image is not symmetric: In this case the two sides are different, but the mirror image of the object is the same, after a rotation by 180° about the axis perpendicular to the mirror plane.
- The sheet surface with the image has a symmetry axis: In this case the two sides are the same, and the mirror image of the object is also the same, again after a rotation by 180° about the axis perpendicular to the mirror plane.

A sheet with a through and through image is achiral. We can distinguish again two cases:

- The sheet surface with the image has no symmetry axis: The two sides are different.
- The sheet surface with the image has a symmetry axis: The two sides are the same.

Configuration Space

The configuration space of a rigid body with one point fixed (i.e., a body with zero translational motion) is given by the underlying manifold of the rotation group SO_3. The configuration space of a nonfixed (with non-zero translational motion) rigid body is E^{+3}, the subgroup of direct isometries of the Euclidean group in three dimensions (combinations of translations and rotations).

Rigid Body Mechanics

In rigid body mechanics we presume that all objects are perfectly rigid. This means they change neither their form nor their volume when forces act on them. It simplifies the following mechanical analysis. Parts of human body are of course not perfectly rigid. Deformations would often make the analysis of motion too complicated. Fluid mechanics is concerned with mechanics of gases and liquids. Relativistic mechanics is related to the Einstein relativity theory and quantum mechanics describes behaviour of objects on atomic and subatomic level. In biomechanics we mostly make use of rigid body mechanics which is best applicable for describing the motion of human body and its parts. Since certain sport events take place in fluid environment, biomechanics also uses knowledge of fluid mechanics. Repeated little deformations of human body may cause injuries, therefore we will also partly inquire into mechanical deformation of human body and its parts.

According to specific approach to the study of objects motion and their equilibrium, mechanics is divided into statics and dynamics. Statics studies objects that are either at rest, or in constant motion, that is a motion with constant velocity as to its magnitude and direction. Dynamics studies objects with acceleration. Dynamics is divided into kinematics and kinetics. Kinematics describes the motion of objects, while kinetics studies forces that cause changes of motion.

Rotational Motion of a Rigid Body

Rotational motion is more complicated than linear motion. A rigid body is an object with a mass that holds a rigid shape, such as a phonograph turntable, in contrast to the sun, which is a ball of gas. Many of the equations for the mechanics of rotating objects are similar to the motion equations for linear motion.

Angular Velocity and Angular Acceleration

The angular displacement of a rotating wheel is the angle between the radius at the beginning and the end of a given time interval. The SI units are radians. The average angular velocity, measured in radians per second, is $\omega = \dfrac{\text{angular displacement}}{\text{elapsed time}} = \dfrac{\theta}{t}$

The angular acceleration has the same form as the linear quantity,

$\alpha = \dfrac{\text{change in angular velocity}}{\text{elapsed time}} = \dfrac{\omega_f - \omega_o}{t}$ and is measured in radians/second/second or rad/s ^{2}.

The kinematics equations for rotational motion at constant angular acceleration are:

$\omega_f = \omega_f \ \alpha t$	Angular velocity as a function of time
$\theta = \frac{1}{2}(\omega_o + \omega_f)t$	Angular displacement as a function of velocity and time
$\theta = \omega_o t + \frac{1}{2}\alpha . t^2$	Angular displacement as a function of time
$\omega_f^{\ 2} = \omega_0^{\ 2} + 2a\theta$	Angular velocity as a function of displacement

Consider a wheel rolling without slipping in a straight line. The forward displacement of the wheel is equal to the linear displacement of a point fixed on the rim.

$d = S = r\theta$

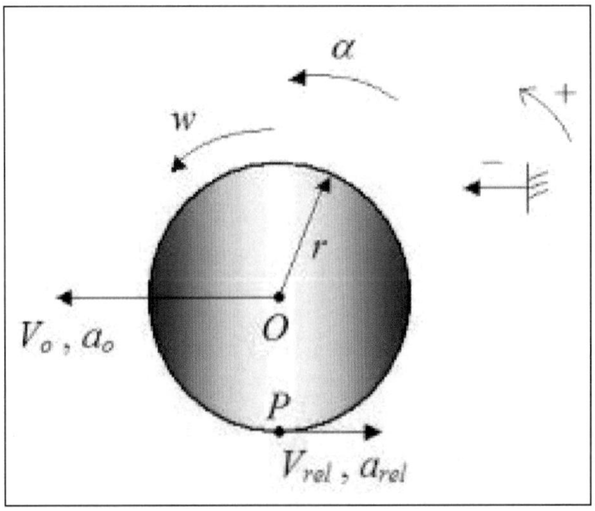

A wheel rolling without slipping.

In this case, the average forward speed of the wheel is $v = d/\,t = (r\theta)/\,t = r\omega$, where r is the distance from the center of rotation to the point of the calculated velocity. The direction of the velocity is tangent to the path of the point of rotation.

The average forward acceleration of the wheel is $a_T = r(\omega f - \omega o)/\,t = r\alpha$. This component of the acceleration is tangential to the point of rotation and represents the changing speed of the object. The direction is the same as the velocity vector.

The radial component of the linear acceleration is $a_r = v^2/r = \omega^2 r$.

Torque

It is easier to open a door by pushing on the edge farthest from the hinges than by pushing in the middle. It is intuitive that the magnitude of the force applied and the distance from the point of application to the hinge affect the tendency of the door to rotate. This physical quantity, torque, is t = r × F sin θ, where *F* is the force applied, *r* is the distance from the point of application to the center of the rotation, and θ is the angle from *r* to *F*.

Moment of Inertia

Substitute Newton's second law into the definition for torque with θ of 90 degrees (a right angle between F and r) and use the relationship between linear acceleration and tangential angular acceleration to obtain $t=rF=rma=mr^2(a/r)=mr^2\alpha$. The quantity mr^2 is defined as moment of inertia of a point mass about the center of rotation.

Imagine two objects of the same mass with different distribution of that mass. The first object might be a heavy ring supported by struts on an axle like a flywheel. The second object could have its mass close to the central axis. Even though the masses of the two objects are equal, it is intuitive that the flywheel will be more difficult to push to a high number of revolutions per second because not only the amount of mass but also the distribution of the mass affects the ease in initiating rotation for a rigid body. The general definition of moment of inertia, also called rotational inertia, for a rigid body is $I = \sum m_i r_i^2$ and is measured in SI units of kilogram-meters².

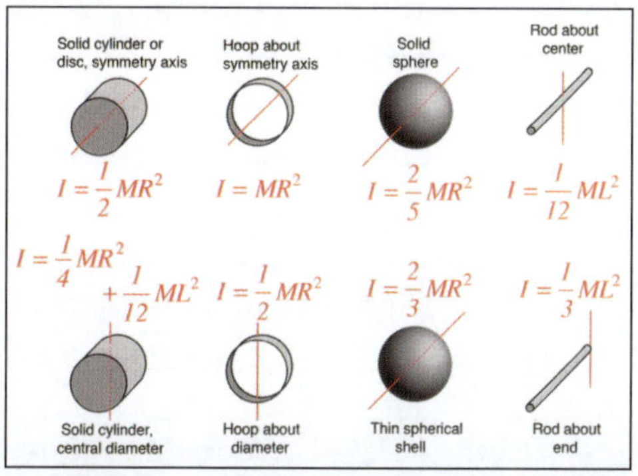

Moments of inertia for various regular shapes.

Mechanics problems frequently include both linear and rotation motions.

Example: In the figure a mass is hanging from a rope wrapped around a pulley. The falling mass *(m)* causes the pulley to rotate, and it is no longer necessary to require the pulley to be massless. Assign mass (*M*) to the pulley and treat it as a rotating disc with radius *(R)*. What is the acceleration of the falling mass, and what is the tension of the rope?

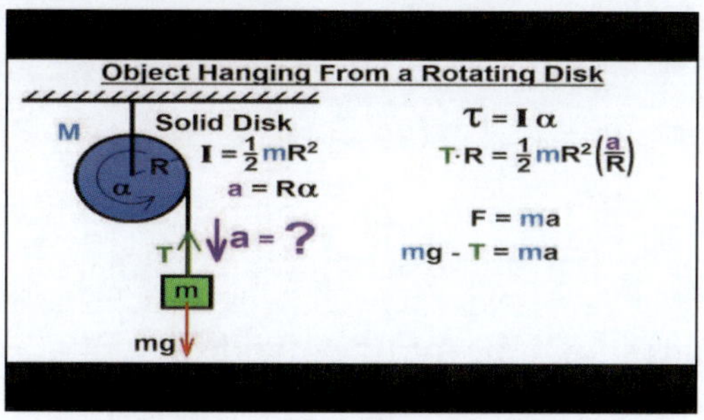

A hanging mass spins a pulley.

The force equation for the falling mass is $T - mg = -ma$. The tension of the rope is the applied force to the edge of the pulley that is causing it to rotate. Thus, $t=I\alpha$, or $TR=(1/2)MR^2(a/R)$, which reduces to $T = (1/2)\,Ma$, where angular acceleration has been replaced by a/R because the cord does not slip and the linear acceleration of the block is equal to the linear acceleration of the rim of the disk. Combining the first and last equation in this example leads to,

Solution: $a = g\dfrac{2m}{M+2m}$ and $T = mg\dfrac{M}{M+2m}$

Angular momentum is rotational momentum that is conserved in the same way that linear momentum is conserved. For a rigid body, the angular momentum *(L)* is the product of the moment of inertia and the angular velocity: $L = I\omega$. For a point of mass, angular momentum can be expressed as the product of linear momentum and the radius (*r*): $L = mvr$. *L* is measured in units of kilograms-meters ^{2} per second or more commonly joule-seconds. The law of conservation of angular momentum can be stated that the angular momentum of a system of objects is conserved if there is not external net torque acting on the system.

Analogous to Newton's law ($F = \Delta(mv)/\Delta t$) there is a rotational counterpart for rotational motion: $t = \Delta L/\Delta t$, or torque is the rate of change of angular momentum.

Consider the example of a child who runs tangential to the edge of a playground merry-go-round with a velocity v_0 and jumps on while the merry-go-round is at rest. The only external forces are that of gravity and the contact forces provided by the support bearings, neither of which causes a torque because they are not applied to cause a horizontal rotation. Treat the child's mass as a point of mass and the merry-go-round as a disc with a radius *R* and mass *M*. From the conservation law, the total angular momentum of the child before the interaction is equal to the total angular momentum of the child and merry-go-round after the collision: $mrv_0 = mrv' + I\omega$, where *r* is the radial distance from the center of the merry-go-round to the place where the child hits. If the child jumps on the edge, *(r = R)* and the angular velocity for the child after the collision can be substituted for the linear velocity, $mRv_0 = mR(R\omega) + (1/2)MR^2$. If the values for the masses and the initial velocity of the child are given, the final velocity of the child and merry-go-round can be calculated.

A single object may have a change in angular velocity due to the conservation of angular momentum if the distribution of the mass of the rigid body is altered. For example, when a figure skater pulls in her extended arms, her moment of inertia will decrease, causing an increase in angular velocity. According to the conservation of angular momentum, $I_o(\omega_o) = I_f(\omega_f)$ where I_o is the moment of inertia of the skater with arms extended, I_f is her moment of inertia with her arms close to her body, ω_0 is her original angular velocity, and ω_f is her final angular velocity.

Rotational kinetic energy, work, and power: Kinetic energy, work, and power are defined in rotational terms as $K.E = (1/2)I\omega^2, W = t\theta, P = t\omega$.

Comparison of dynamics equation for linear and rotational motion: The dynamic relations are given to compare the equation for linear and rotational motion.

Table: Comparison of dynamics equations for linear and rotational motion.

	Linear Motion	Rotational Motion
Newton's second law	F = ma	$t = Ia$
Momentum	p = mv	$L = I\omega$
Work	$W = F\Delta x$ or $W = \int F \cdot dx$	$W = t\Delta\theta$ or $W = \int t \cdot d\theta$

Kinetic energy	$K \cdot E \cdot \frac{1}{2} mv^2$	$K \cdot E \cdot \frac{1}{2} I\omega^2$
Power	P = Fv	$P = t\omega$

Rigid Body Dynamics

Rigid body dynamics studies the movement of systems of interconnected bodies under the action of external forces. The assumption that the bodies are rigid, which means that they do not deform under the action of applied forces, simplifies the analysis by reducing the parameters that describe the configuration of the system to the translation and rotation of reference frames attached to each body. This excludes bodies that display fluid, highly elastic, and plastic behavior.

The dynamics of a rigid body system is described by the laws of kinematics and by the application of Newton's second law (kinetics) or their derivative form Lagrangian mechanics. The solution of these equations of motion provides a description of the position, the motion and the acceleration of the individual components of the system and overall the system itself, as a function of time. The formulation and solution of rigid body dynamics is an important tool in the computer simulation of mechanical systems.

Planar Rigid Body Dynamics

If a system of particles moves parallel to a fixed plane, the system is said to be constrained to planar movement. In this case, Newton's laws (kinetics) for a rigid system of N particles, P_i, i=1,...,N, simplify because there is no movement in the *k* direction. Determine the resultant force and torque at a reference point R, to obtain,

$$\mathbf{F} = \sum_{i=1}^{N} m_i \mathbf{A}_i, \quad \mathbf{T} = \sum_{i=1}^{N} (\mathbf{r}_i - \mathbf{R}) \times (m_i \mathbf{A}_i),$$

where r_i denotes the planar trajectory of each particle.

The kinematics of a rigid body yields the formula for the acceleration of the particle P_i in terms of the position R and acceleration A of the reference particle as well as the angular velocity vector ω and angular acceleration vector α of the rigid system of particles as,

$$\mathbf{A}_i = \alpha \times (\mathbf{r}_i - \mathbf{R}) + \omega \times (\omega \times (\mathbf{r}_i - \mathbf{R})) + \mathbf{A}$$

For systems that are constrained to planar movement, the angular velocity and angular acceleration vectors are directed along *k* perpendicular to the plane of movement, which simplifies this acceleration equation. In this case, the acceleration vectors can be simplified by introducing the unit vectors e_i from the reference point R to a point r_i and the unit vectors $\mathbf{t}_i = k \times \mathbf{e}_i$, so

$$\mathbf{A}_i = \alpha(\Delta r_i \mathbf{t}_i) - \omega^2 (\Delta r_i \mathbf{e}_i) + \mathbf{A}$$

This yields the resultant force on the system as:

$$\mathbf{F} = \alpha \sum_{i=1}^{N} m_i (\Delta r_i \mathbf{t}_i) - \omega^2 \sum_{i=1}^{N} m_i (\Delta r_i \mathbf{e}_i) + (\sum_{i=1}^{N} m_i)\mathbf{A},$$

and torque as,

$$\mathbf{T} = \sum_{i=1}^{N} (m_i \Delta r_i \mathbf{e}_i) \times (\alpha(\Delta r_i \mathbf{t}_i) - \omega^2 (\Delta r_i \mathbf{e}_i) + \mathbf{A}) = (\sum_{i=1}^{N} m_i \Delta r_i^2)\alpha \vec{k} + (\sum_{i=1}^{N} m_i \Delta r_i \mathbf{e}_i) \times \mathbf{A},$$

Where $\mathbf{e}_i \times \mathbf{e}_i = 0$ and $\mathbf{e}_i \times \mathbf{t}_i = k$ is the unit vector perpendicular to the plane for all of the particles P_i.

Use the center of mass C as the reference point, so these equations for Newton's laws simplify to become,

$$\mathbf{F} = M\mathbf{A}, \quad \mathbf{T} = I_C \alpha \vec{k},$$

where M is the total mass and I_C is the moment of inertia about an axis perpendicular to the movement of the rigid system and through the center of mass.

Rigid Body in Three Dimensions

Newton's Second Law in Three Dimensions

To consider rigid body dynamics in three-dimensional space, Newton's second law must be extended to define the relationship between the movement of a rigid body and the system of forces and torques that act on it.

Newton formulated his second law for a particle as, "The change of motion of an object is proportional to the force impressed and is made in the direction of the straight line in which the force is impressed." Because Newton generally referred to mass times velocity as the "motion" of a particle, the phrase "change of motion" refers to the mass times acceleration of the particle, and so this law is usually written as,

$$\mathbf{F} = m\mathbf{a},$$

where F is understood to be the only external force acting on the particle, *m* is the mass of the particle, and a is its acceleration vector. The extension of Newton's second law to rigid bodies is achieved by considering a rigid system of particles.

Rigid System of Particles

If a system of *N* particles, P_i, i = 1,...,*N*, are assembled into a rigid body, then Newton's second law can be applied to each of the particles in the body. If F_i is the external force applied to particle P_i with mass m_i, then,

$$\mathbf{F}_i + \sum_{j=1}^{N} \mathbf{F}_{ij} = m_i \mathbf{a}_i, \quad i = 1, \ldots, N,$$

where F_{ij} is the internal force of particle P_j acting on particle P_i that maintains the constant distance between these particles.

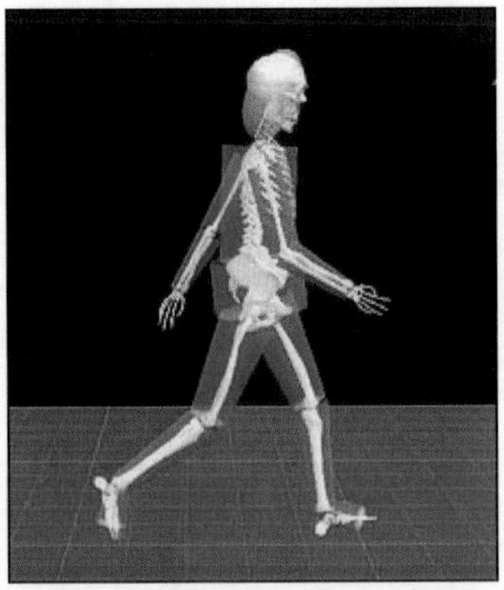

Human body modelled as a system of rigid bodies of geometrical solids.
Representative bones were added for better visualization of the walking person.

An important simplification to these force equations is obtained by introducing the resultant force and torque that acts on the rigid system. This resultant force and torque is obtained by choosing one of the particles in the system as a reference point, R, where each of the external forces are applied with the addition of an associated torque. The resultant force F and torque T are given by the formulas,

$$\mathbf{F} = \sum_{i=1}^{N} \mathbf{F}_i, \quad \mathbf{T} = \sum_{i=1}^{N} (\mathbf{R}_i - \mathbf{R}) \times \mathbf{F}_i,$$

where R_i is the vector that defines the position of particle P_i.

Newton's second law for a particle combines with these formulas for the resultant force and torque to yield,

$$\mathbf{F} = \sum_{i=1}^{N} m_i \mathbf{a}_i, \quad \mathbf{T} = \sum_{i=1}^{N} (\mathbf{R}_i - \mathbf{R}) \times (m_i \mathbf{a}_i),$$

where the internal forces F_{ij} cancel in pairs. The kinematics of a rigid body yields the formula for the acceleration of the particle P_i in terms of the position R and acceleration a of the reference particle as well as the angular velocity vector ω and angular acceleration vector α of the rigid system of particles as,

$$\mathbf{a}_i = \alpha \times (\mathbf{R}_i - \mathbf{R}) + \omega \times (\omega \times (\mathbf{R}_i - \mathbf{R})) + \mathbf{a}$$

Mass Properties

The mass properties of the rigid body are represented by its center of mass and inertia matrix. Choose the reference point R so that it satisfies the condition,

$$\sum_{i=1}^{N} m_i (\mathbf{R}_i - \mathbf{R}) = 0,$$

then it is known as the center of mass of the system. The inertia matrix $[I_R]$ of the system relative to the reference point R is defined by:

$$[I_R] = \sum_{i=1}^{N} m_i (\mathbf{I}(\mathbf{S}_i^T \mathbf{S}_i) - \mathbf{S}_i \mathbf{S}_i^T),$$

where $\mathbf{S}_i$ is the column vector $R_i - R$; and $\mathbf{S}_i^T$ is its transpose.

$\mathbf{S}_i^T \mathbf{S}_i$ is the scalar product of $\mathbf{S}_i$ with itself, while $\mathbf{S}_i \mathbf{S}_i^T$ is the tensor product of $\mathbf{S}_i$ with itself.

I is the 3 by 3 identity matrix.

Force-torque Equations

Using the center of mass and inertia matrix, the force and torque equations for a single rigid body take the form:

$$\mathbf{F} = m\mathbf{a}, \quad \mathbf{T} = [I_R]\alpha + \omega \times [I_R]\omega,$$

and are known as Newton's second law of motion for a rigid body.

The dynamics of an interconnected system of rigid bodies, $Bi, j = 1, \ldots, M$, is formulated by isolating each rigid body and introducing the interaction forces. The resultant of the external and interaction forces on each body, yields the force-torque equations:

$$\mathbf{F}_j = m_j \mathbf{a}_j, \quad \mathbf{T}_j = [I_R]_j \alpha_j + \omega_j \times [I_R]_j \omega_j, \quad j = 1, \ldots, M.$$

Newton's formulation yields 6M equations that define the dynamics of a system of M rigid bodies.

Rotation in Three Dimensions

A rotating object, whether under the influence of torques or not, may exhibit the behaviours of precession and nutation. The fundamental equation describing the behavior of a rotating solid body is Euler's equation of motion:

$$\tau = \frac{D\mathbf{L}}{Dt} = \frac{d\mathbf{L}}{dt} + \omega \times \mathbf{L} = \frac{d(I\omega)}{dt} + \omega \times I\omega = I\alpha + \omega \times I\omega$$

Where the pseudovectors τ and L are, respectively, the torques on the body and its angular momentum, the scalar I is its moment of inertia, the vector ω is its angular velocity, the vector α is its angular acceleration, D is the differential in an inertial reference frame and d is the differential in a relative reference frame fixed with the body.

It follows from Euler's equation that a torque τ applied perpendicular to the axis of rotation, and therefore perpendicular to L, results in a rotation about an axis perpendicular to both τ and L. This motion is called *precession*. The angular velocity of precession Ω_P is given by the cross product:

$$\tau = \Omega_P \times \mathbf{L}.$$

Precession can be demonstrated by placing a spinning top with its axis horizontal and supported loosely (frictionless toward precession) at one end. Instead of falling, as might be expected, the top appears to defy gravity by remaining with its axis horizontal, when the other end of the axis is left unsupported and the free end of the axis slowly describes a circle in a horizontal plane, the resulting precession turning. This effect is explained by the above equations. The torque on the top is supplied by a couple of forces: gravity acting downward on the device's centre of mass, and an equal force acting upward to support one end of the device. The rotation resulting from this torque is not downward, as might be intuitively expected, causing the device to fall, but perpendicular to both the gravitational torque (horizontal and perpendicular to the axis of rotation) and the axis of rotation (horizontal and outwards from the point of support), i.e., about a vertical axis, causing the device to rotate slowly about the supporting point.

Under a constant torque of magnitude τ, the speed of precession Ω_P is inversely proportional to L, the magnitude of its angular momentum:

$$\tau = \Omega_P L \sin\theta,$$

Where θ is the angle between the vectors Ω_P and L. Thus, if the top's spin slows down (for example, due to friction), its angular momentum decreases and so the rate of precession increases. This continues until the device is unable to rotate fast enough to support its own weight, when it stops precessing and falls off its support, mostly because friction against precession cause another precession that goes to cause the fall.

By convention, these three vectors - torque, spin, and precession - are all oriented with respect to each other according to the right-hand rule.

Virtual Work of Forces Acting on a Rigid Body

An alternate formulation of rigid body dynamics that has a number of convenient features is obtained by considering the virtual work of forces acting on a rigid body.

The virtual work of forces acting at various points on a single rigid body can be calculated using the velocities of their point of application and the resultant force and torque. To see this, let the forces F_1, F_2... Fn act on the points R_1, R_2... Rn in a rigid body.

The trajectories of Ri, $i = 1, ..., n$ are defined by the movement of the rigid body. The velocity of the points Ri along their trajectories are,

$$\mathbf{V}_i = \vec{\omega} \times (\mathbf{R}_i - \mathbf{R}) + \mathbf{V},$$

where ω is the angular velocity vector of the body.

Virtual Work

Work is computed from the dot product of each force with the displacement of its point of contact,

$$\delta W = \sum_{i=1}^{n} \mathbf{F}_i \cdot \delta \mathbf{r}_i.$$

If the trajectory of a rigid body is defined by a set of generalized coordinates $qj, j = 1, \ldots, m$, then the virtual displacements δr_i are given by,

$$\delta \mathbf{r}_i = \sum_{j=1}^{m} \frac{\partial \mathbf{r}_i}{\partial q_j} \delta q_j = \sum_{j=1}^{m} \frac{\partial \mathbf{V}_i}{\partial \dot{q}_j} \delta q_j.$$

The virtual work of this system of forces acting on the body in terms of the generalized coordinates becomes,

$$\delta W = \mathbf{F}_1 \cdot \left(\sum_{j=1}^{m} \frac{\partial \mathbf{V}_1}{\partial \dot{q}_j} \delta q_j \right) + \ldots + \mathbf{F}_n \cdot \left(\sum_{j=1}^{m} \frac{\partial \mathbf{V}_n}{\partial \dot{q}_j} \delta q_j \right)$$

or collecting the coefficients of δq_j,

$$\delta W = \left(\sum_{i=1}^{n} \mathbf{F}_i \cdot \frac{\partial \mathbf{V}_i}{\partial \dot{q}_1} \right) \delta q_1 + \ldots + \left(\sum_{1=1}^{n} \mathbf{F}_i \cdot \frac{\partial \mathbf{V}_i}{\partial \dot{q}_m} \right) \delta q_m.$$

Generalized Forces

For simplicity consider a trajectory of a rigid body that is specified by a single generalized coordinate q, such as a rotation angle, then the formula becomes,

$$\delta W = \left(\sum_{i=1}^{n} \mathbf{F}_i \cdot \frac{\partial \mathbf{V}_i}{\partial \dot{q}} \right) \delta q = \left(\sum_{i=1}^{n} \mathbf{F}_i \cdot \frac{\partial (\vec{\omega} \times (\mathbf{R}_i - \mathbf{R}) + \mathbf{V})}{\partial \dot{q}} \right) \delta q.$$

Introduce the resultant force F and torque T so this equation takes the form,

$$\delta W = \left(\mathbf{F} \cdot \frac{\partial \mathbf{V}}{\partial \dot{q}} + \mathbf{T} \cdot \frac{\partial \vec{\omega}}{\partial \dot{q}} \right) \delta q.$$

The quantity Q defined by,

$$Q = \mathbf{F} \cdot \frac{\partial \mathbf{V}}{\partial \dot{q}} + \mathbf{T} \cdot \frac{\partial \vec{\omega}}{\partial \dot{q}},$$

is known as the generalized force associated with the virtual displacement δq. This formula generalizes to the movement of a rigid body defined by more than one generalized coordinate, that is,

$$\delta W = \sum_{j=1}^{m} Q_j \delta q_j,$$

Where,

$$Q_j = \mathbf{F} \cdot \frac{\partial \mathbf{V}}{\partial \dot{q}_j} + \mathbf{T} \cdot \frac{\partial \vec{\omega}}{\partial \dot{q}_j}, \quad j = 1, \ldots, m.$$

It is useful to note that conservative forces such as gravity and spring forces are derivable from a potential function $V(q_1,\ldots, qn)$, known as a potential energy. In this case the generalized forces are given by

$$Q_j = -\frac{\partial V}{\partial q_j}, \quad j = 1,\ldots,m.$$

D'Alembert's form of the Principle of Virtual Work

The equations of motion for a mechanical system of rigid bodies can be determined using D'Alembert's form of the principle of virtual work. The principle of virtual work is used to study the static equilibrium of a system of rigid bodies, however by introducing acceleration terms in Newton's laws this approach is generalized to define dynamic equilibrium.

Static Equilibrium

The static equilibrium of a mechanical system rigid bodies is defined by the condition that the virtual work of the applied forces is zero for any virtual displacement of the system. This is known as the principle of virtual work. This is equivalent to the requirement that the generalized forces for any virtual displacement are zero, that is $Q_i = 0$.

Let a mechanical system be constructed from n rigid bodies, B_i, i = 1,...,n, and let the resultant of the applied forces on each body be the force-torque pairs, F_i and T_i, i = 1,...,n. Notice that these applied forces do not include the reaction forces where the bodies are connected. Finally, assume that the velocity V_i and angular velocities ω_i, i =,1...,n, for each rigid body, are defined by a single generalized coordinate q. Such a system of rigid bodies is said to have one degree of freedom.

The virtual work of the forces and torques, F_i and T_i, applied to this one degree of freedom system is given by:

$$\delta W = \sum_{i=1}^{n} (\mathbf{F}_i \cdot \frac{\partial \mathbf{V}_i}{\partial \dot{q}} + \mathbf{T}_i \cdot \frac{\partial \vec{\omega}_i}{\partial \dot{q}}) \delta q = Q \delta q,$$

Where,

$$Q = \sum_{i=1}^{n} (\mathbf{F}_i \cdot \frac{\partial \mathbf{V}_i}{\partial \dot{q}} + \mathbf{T}_i \cdot \frac{\partial \vec{\omega}_i}{\partial \dot{q}}),$$

is the generalized force acting on this one degree of freedom system.

If the mechanical system is defined by m generalized coordinates, q_j, j=1,...,m, then the system has m degrees of freedom and the virtual work is given by,

$$\delta W = \sum_{j=1}^{m} Q_j \delta q_j,$$

Where,

$$Q_j = \sum_{i=1}^{n} (\mathbf{F}_i \cdot \frac{\partial \mathbf{V}_i}{\partial \dot{q}_j} + \mathbf{T}_i \cdot \frac{\partial \vec{\omega}_i}{\partial \dot{q}_j}), \quad j = 1, \ldots, m.$$

is the generalized force associated with the generalized coordinate q_j. The principle of virtual work states that static equilibrium occurs when these generalized forces acting on the system are zero, that is,

$$Q_j = 0, \quad j = 1, \ldots, m.$$

These m equations define the static equilibrium of the system of rigid bodies.

Generalized Inertia Forces

Consider a single rigid body which moves under the action of a resultant force F and torque T, with one degree of freedom defined by the generalized coordinate q. Assume the reference point for the resultant force and torque is the center of mass of the body, then the generalized inertia force Q* associated with the generalized coordinate q is given by:

$$Q^* = -(M\mathbf{A}) \cdot \frac{\partial \mathbf{V}}{\partial \dot{q}} - ([I_R]\alpha + \omega \times [I_R]\omega) \cdot \frac{\partial \vec{\omega}}{\partial \dot{q}}.$$

This inertia force can be computed from the kinetic energy of the rigid body,

$$T = \frac{1}{2} M\mathbf{V} \cdot \mathbf{V} + \frac{1}{2} \vec{\omega} \cdot [I_R] \vec{\omega},$$

by using the formula,

$$Q^* = -\left(\frac{d}{dt} \frac{\partial T}{\partial \dot{q}} - \frac{\partial T}{\partial q} \right).$$

A system of n rigid bodies with m generalized coordinates has the kinetic energy:

$$T = \sum_{i=1}^{n} (\frac{1}{2} M\mathbf{V}_i \cdot \mathbf{V}_i + \frac{1}{2} \vec{\omega}_i \cdot [I_R] \vec{\omega}_i),$$

which can be used to calculate the m generalized inertia forces,

$$Q_j^* = -\left(\frac{d}{dt} \frac{\partial T}{\partial \dot{q}_j} - \frac{\partial T}{\partial q_j} \right), \quad j = 1, \ldots, m.$$

Dynamic Equilibrium

D'Alembert's form of the principle of virtual work states that a system of rigid bodies is in dynamic equilibrium when the virtual work of the sum of the applied forces and the inertial forces is zero for any virtual displacement of the system. Thus, dynamic equilibrium of a system of n rigid bodies with m generalized coordinates requires that,

$$\delta W = (Q_1 + Q_1^*)\delta q_1 + \ldots + (Q_m + Q_m^*)\delta q_m = 0,$$

for any set of virtual displacements δq_j. This condition yields m equations,

$$Q_j + Q_j^* = 0, \quad j = 1, \ldots, m,$$

which can also be written as,

$$\frac{d}{dt}\frac{\partial T}{\partial \dot{q}_j} - \frac{\partial T}{\partial q_j} = Q_j, \quad j = 1, \ldots, m.$$

The result is a set of m equations of motion that define the dynamics of the rigid body system.

Lagrange's Equations

If the generalized forces Q_j are derivable from a potential energy $V(q_1,\ldots,q_m)$, then these equations of motion take the form,

$$\frac{d}{dt}\frac{\partial T}{\partial \dot{q}_j} - \frac{\partial T}{\partial q_j} = -\frac{\partial V}{\partial q_j}, \quad j = 1, \ldots, m.$$

In this case, introduce the Lagrangian, L=T-V, so these equations of motion become,

$$\frac{d}{dt}\frac{\partial L}{\partial \dot{q}_j} - \frac{\partial L}{\partial q_j} = 0 \quad j = 1, \ldots, m.$$

These are known as Lagrange's equations of motion.

Linear and Angular Momentum

System of Particles

The linear and angular momentum of a rigid system of particles is formulated by measuring the position and velocity of the particles relative to the center of mass. Let the system of particles P_i, i=1,...,n be located at the coordinates r_i and velocities v_i. Select a reference point R and compute the relative position and velocity vectors,

$$\mathbf{r}_i = (\mathbf{r}_i - \mathbf{R}) + \mathbf{R}, \quad \mathbf{v}_i = \frac{d}{dt}(\mathbf{r}_i - \mathbf{R}) + \mathbf{V}.$$

The total linear and angular momentum vectors relative to the reference point R are:

$$\mathbf{p} = \frac{d}{dt}\left(\sum_{i=1}^{n} m_i(\mathbf{r}_i - \mathbf{R})\right) + \left(\sum_{i=1}^{n} m_i\right)\mathbf{V},$$

and,

$$\mathbf{L} = \sum_{i=1}^{n} m_i(\mathbf{r}_i - \mathbf{R}) \times \frac{d}{dt}(\mathbf{r}_i - \mathbf{R}) + (\sum_{i=1}^{n} m_i(\mathbf{r}_i - \mathbf{R})) \times \mathbf{V}.$$

If R is chosen as the center of mass these equations simplify to:

$$\mathbf{p} = M\mathbf{V}, \quad \mathbf{L} = \sum_{i=1}^{n} m_i(\mathbf{r}_i - \mathbf{R}) \times \frac{d}{dt}(\mathbf{r}_i - \mathbf{R}).$$

Rigid System of Particles

To specialize these formulas to a rigid body, assume the particles are rigidly connected to each other so P_i, i = 1,...,n are located by the coordinates r_i and velocities v_i. Select a reference point R and compute the relative position and velocity vectors,

$$\mathbf{r}_i = (\mathbf{r}_i - \mathbf{R}) + \mathbf{R}, \quad \mathbf{v}_i = \omega \times (\mathbf{r}_i - \mathbf{R}) + \mathbf{V},$$

where ω is the angular velocity of the system.

The linear momentum and angular momentum of this rigid system measured relative to the center of mass R is,

$$\mathbf{p} = (\sum_{i=1}^{n} m_i)\mathbf{V}, \quad \mathbf{L} = \sum_{i=1}^{n} m_i(\mathbf{r}_i - \mathbf{R}) \times \mathbf{v}_i = \sum_{i=1}^{n} m_i(\mathbf{r}_i - \mathbf{R}) \times (\omega \times (\mathbf{r}_i - \mathbf{R}))$$

These equations simplify to become,

$$\mathbf{p} = M\mathbf{V}, \quad \mathbf{L} = [I_R]\omega,$$

where M is the total mass of the system and $[I_R]$ is the moment of inertia matrix defined by:

$$[I_R] = -\sum_{i=1}^{n} m_i [r_i - R][r_i - R],$$

where $[r_i\text{-}R]$ is the skew-symmetric matrix constructed from the vector $r_i\text{-}R$.

Rigid Line Inclusion

A rigid line inclusion, also called stiffener, is a mathematical model used in solid mechanics to describe a narrow hard phase, dispersed within a matrix material. This inclusion is idealised as an infinitely rigid and thin reinforcement, so that it represents a sort of 'inverse' crack, from which the nomenclature 'anticrack' derives.

From the mechanical point of view, a stiffener introduces a kinematical constraint, imposing that it may only suffer a rigid body motion along its line.

Theoretical Model

The stiffener model has been used to investigate different mechanical problems in classical elasticity (load diffusion, inclusion at bi material interface).

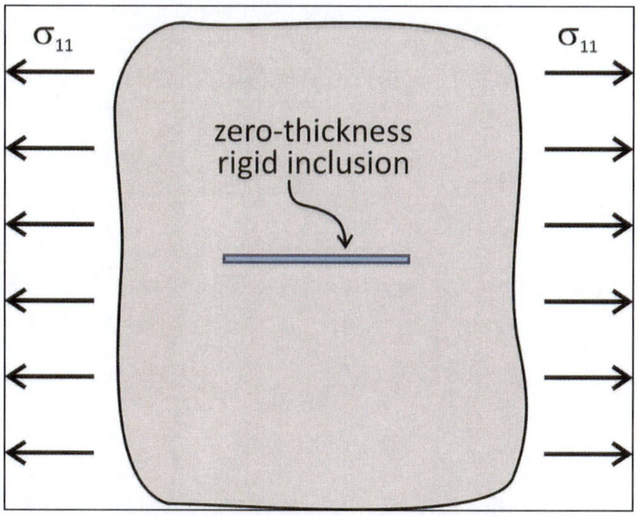

Sketch of a stiffener embedded in a matrix loaded at its boundary.

The main characteristics of the theoretical solutions are basically the following:

Similarly to a fracture, a square-root singularity in the stress/strain fields is present at the tip of the inclusion.

- In a homogeneous matrix subject to uniform stress at infinity, such singularity only arises when a normal stress acts parallel or orthogonal to the inclusion line, while a stiffener parallel to a simple shear does not disturb the ambient field.

Experimental Validation

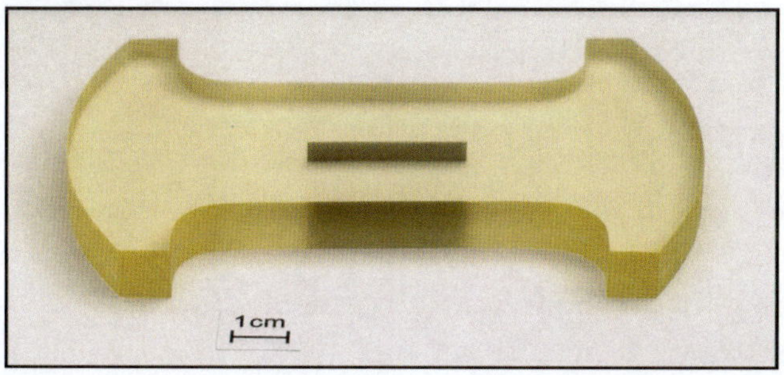

Dog-bone shaped sample of two-component epoxy resin containing a lamellar (aluminum) inclusion.

The characteristics of the elastic solution have been experimentally confirmed through photoelastic transmission experiments.

Shear Bands Emerging at the Stiffener Tip

Analytical solutions obtained in prestressed elasticity show the possibility of the emergence of shear bands at the tip of the stiffener.

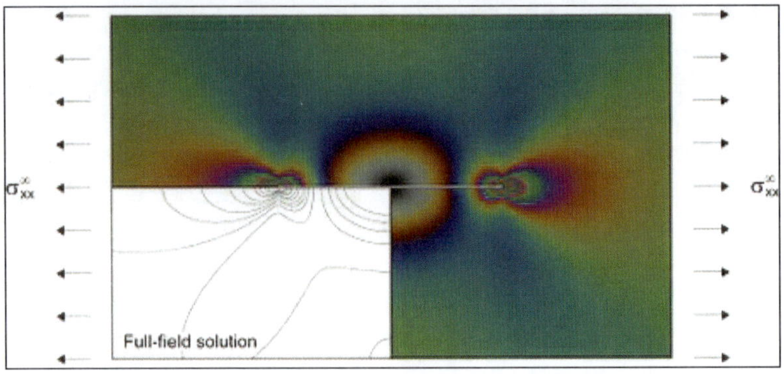

Photoelastic experiment to validate the rigid line inclusion model. Isochromatic fringe patterns around a steel platelet in a photo-elastic two-part epoxy resin compared to analytical solution obtained in plane-strain classical elasticity. Normal stress parallel to the inclusion line is applied.

Rigid Rotor

The rigid rotor is a mechanical model of rotating systems. An arbitrary rigid rotor is a 3-dimensional rigid object, such as a top. To orient such an object in space requires three angles, known as Euler angles. A special rigid rotor is the linear rotor requiring only two angles to describe, for example of a diatomic molecule. More general molecules are 3-dimensional, such as water (asymmetric rotor), ammonia (symmetric rotor), or methane (spherical rotor).

Linear Rotor

The linear rigid rotor model consists of two point masses located at fixed distances from their center of mass. The fixed distance between the two masses and the values of the masses are the only characteristics of the rigid model. However, for many actual diatomics this model is too restrictive since distances are usually not completely fixed. Corrections on the rigid model can be made to compensate for small variations in the distance. Even in such a case the rigid rotor model is a useful point of departure (zeroth-order model).

Classical Linear Rigid Rotor

The classical linear rotor consists of two point masses m_1 and m_2 (with reduced mass $\mu = \frac{m_1 m_2}{m_1 + m_2}$) each at a distance R. The rotor is rigid if R is independent of time. The kinematics of a linear rigid rotor is usually described by means of spherical polar coordinates, which form a coordinate system of R^3. In the physics convention the coordinates are the co-latitude (zenith) angle θ, the longitudinal (azimuth) angle φ and the distance R. The angles specify the orientation of the rotor in space. The kinetic energy T of the linear rigid rotor is given by:

$$2T = \mu R^2\left[\dot{\theta}^2 + (\dot{\varphi}\sin\theta)^2\right] = \mu R^2\begin{pmatrix}\dot{\theta} & \dot{\varphi}\end{pmatrix}\begin{pmatrix}1 & 0\\ 0 & \sin^2\theta\end{pmatrix}\begin{pmatrix}\dot{\theta}\\ \dot{\varphi}\end{pmatrix} = \mu\begin{pmatrix}\dot{\theta} & \dot{\varphi}\end{pmatrix}\begin{pmatrix}h_\theta^2 & 0\\ 0 & h_\varphi^2\end{pmatrix}\begin{pmatrix}\dot{\theta}\\ \dot{\varphi}\end{pmatrix},$$

Where, $h_\theta = R$ and $h_\varphi = R\sin\theta$ are scale (or Lamé) factors.

Scale factors are of importance for quantum mechanical applications since they enter the Laplacian expressed in curvilinear coordinates. In the case at hand (constant R):

$$\nabla^2 = \frac{1}{h_\theta h_\varphi}\left[\frac{\partial}{\partial\theta}\frac{h_\varphi}{h_\theta}\frac{\partial}{\partial\theta} + \frac{\partial}{\partial\varphi}\frac{h_\theta}{h_\varphi}\frac{\partial}{\partial\varphi}\right] = \frac{1}{R^2}\left[\frac{1}{\sin\theta}\frac{\partial}{\partial\theta}\sin\theta\frac{\partial}{\partial\theta} + \frac{1}{\sin^2\theta}\frac{\partial^2}{\partial\varphi^2}\right].$$

The classical Hamiltonian function of the linear rigid rotor is:

$$H = \frac{1}{2\mu R^2}\left[p_\theta^2 + \frac{p_\varphi^2}{\sin^2\theta}\right].$$

Quantum Mechanical Linear Rigid Rotor

The linear rigid rotor model can be used in quantum mechanics to predict the rotational energy of a diatomic molecule. The rotational energy depends on the moment of inertia for the system, I. In the center of mass reference frame, the moment of inertia is equal to:

$$I = \mu R^2$$

Where, μ is the reduced mass of the molecule and R is the distance between the two atoms.

According to quantum mechanics, the energy levels of a system can be determined by solving the Schrödinger equation:

$$\hat{H}\Psi = E\Psi$$

Where, ψ is the wave function and $\hat{H}$ is the energy (Hamiltonian) operator. For the rigid rotor in a field-free space, the energy operator corresponds to the kinetic energy of the system:

$$\hat{H} = -\frac{\hbar^2}{2\mu}\nabla^2$$

Where, $\hbar$ is reduced Planck constant and ∇^2 is the Laplacian. The Laplacian is given above in terms of spherical polar coordinates. The energy operator written in terms of these coordinates is:

$$\hat{H} = -\frac{\hbar^2}{2I}\left[\frac{1}{\sin\theta}\frac{\partial}{\partial\theta}\left(\sin\theta\frac{\partial}{\partial\theta}\right) + \frac{1}{\sin^2\theta}\frac{\partial^2}{\partial\varphi^2}\right]$$

This operator appears also in the Schrödinger equation of the hydrogen atom after the radial part is separated off. The eigenvalue equation becomes:

$$\hat{H}Y_\ell^m(\theta,\varphi) = \frac{\hbar^2}{2I}\ell(\ell+1)Y_\ell^m(\theta,\varphi)$$

The symbol $Y_\ell^m(\theta,\varphi)$ represents a set of functions known as the spherical harmonics. Note that the energy does not depend on m. The energy,

$$E_{\ell} = \frac{\hbar^2}{2I} \ell(\ell+1)$$

is $2\ell+1$-fold degenerate: the functions with fixed ℓ and $m = -\ell, -\ell+1, \ldots, \ell$ have the same energy.

Introducing the rotational constant B, we write,

$$E_{\ell} = B\,\ell(\ell+1) \quad \text{with} \quad B \equiv \frac{\hbar^2}{2I}.$$

In the units of reciprocal length the rotational constant is,

$$\bar{B} \equiv \frac{B}{hc} = \frac{h}{8\pi^2 cI} = \frac{\hbar}{4\pi c\mu R_e^2},$$

with c the speed of light. If cgs units are used for h, c, and I, $\bar{B}$ is expressed in wave numbers, cm^{-1}, a unit that is often used for rotational-vibrational spectroscopy. The rotational constant $\bar{B}(R)$ depends on the distance R. Often one writes $B_e = \bar{B}(R_e)$ where R_e is the equilibrium value of R (the value for which the interaction energy of the atoms in the rotor has a minimum).

A typical rotational spectrum consists of a series of peaks that correspond to transitions between levels with different values of the angular momentum quantum number (ℓ). Consequently, rotational peaks appear at energies corresponding to an integer multiple of $2\bar{B}$.

Selection Rules

Rotational transitions of a molecule occur when the molecule absorbs a photon [a particle of a quantized electromagnetic (em) field]. Depending on the energy of the photon (i.e., the wavelength of the em field) this transition may be seen as a sideband of a vibrational and electronic transition. Pure rotational transitions, in which the vibronic (= vibrational plus electronic) wave function does not change, occur in the microwave region of the electromagnetic spectrum.

Typically, rotational transitions can only be observed when the angular momentum quantum number changes by 1 ($\Delta l = \pm 1$). This selection rule arises from a first-order perturbation theory approximation of the time-dependent Schrödinger equation. According to this treatment, rotational transitions can only be observed when one or more components of the dipole operator have a non-vanishing transition moment. If z is the direction of the electric field component of the incoming electromagnetic wave, the transition moment is,

$$\langle \psi_2 | \mu_z | \psi_1 \rangle = (\mu_z)_{21} = \int \psi_2^* \mu_z \psi_1 \, \mathrm{d}\tau.$$

A transition occurs if this integral is non-zero. By separating the rotational part of the molecular wavefunction from the vibronic part, one can show that this means that the molecule must have a permanent dipole moment. After integration over the vibronic coordinates the following rotational part of the transition moment remains,

$$(\mu_z)_{l,m;l',m'} = \mu \int_0^{2\pi} \mathrm{d}\phi \int_0^{\pi} Y_{l'}^{m'}(\theta,\phi)^* \cos\theta \, Y_l^m(\theta,\phi) \, \mathrm{d}\cos\theta.$$

Here, $\mu \cos\theta$ is the z component of the permanent dipole moment. The moment μ is the vibronically averaged component of the dipole operator. Only the component of the permanent dipole along the axis of a heteronuclear molecule is non-vanishing. By the use of the orthogonality of the spherical harmonics $Y_l^m(\theta,\phi)$ it is possible to determine which values of l, m, l' and m' will result in nonzero values for the dipole transition moment integral. This constraint results in the observed selection rules for the rigid rotor:

$$\Delta m = 0 \quad \text{and} \quad \Delta l = \pm 1$$

Non-rigid Linear Rotor

The rigid rotor is commonly used to describe the rotational energy of diatomic molecules but it is not a completely accurate description of such molecules. This is because molecular bonds (and therefore the interatomic distance R) are not completely fixed; the bond between the atoms stretches out as the molecule rotates faster (higher values of the rotational quantum number l). This effect can be accounted for by introducing a correction factor known as the centrifugal distortion constant $\bar{D}$ (bars on top of various quantities indicate that these quantities are expressed in cm^{-1}):

$$\bar{E}_l = \frac{E_l}{hc} = \bar{B}l(l+1) - \bar{D}l^2(l+1)^2$$

Where,

$$\bar{D} = \frac{4\bar{B}^3}{\bar{\omega}^2}$$

$\bar{\omega}$ is the fundamental vibrational frequency of the bond (in cm^{-1}). This frequency is related to the reduced mass and the force constant (bond strength) of the molecule according to,

$$\bar{\omega} = \frac{1}{2\pi c}\sqrt{\frac{k}{\mu}}$$

The non-rigid rotor is an acceptably accurate model for diatomic molecules but is still somewhat imperfect. This is because, although the model does account for bond stretching due to rotation, it ignores any bond stretching due to vibrational energy in the bond (anharmonicity in the potential).

Arbitrarily Shaped Rigid Rotor

An arbitrarily shaped rigid rotor is a rigid body of arbitrary shape with its center of mass fixed (or in uniform rectilinear motion) in field-free space R^3, so that its energy consists only of rotational kinetic energy (and possibly constant translational energy that can be ignored). A rigid body can be (partially) characterized by the three eigenvalues of its moment of inertia tensor, which are real nonnegative values known as *principal moments of inertia*. In microwave spectroscopy—the spectroscopy based on rotational transitions—one usually classifies molecules (seen as rigid rotors) as follows:

- Spherical rotors.

- Symmetric rotors:
  - Oblate symmetric rotors.
  - Prolate symmetric rotors.
- Asymmetric rotors.

This classification depends on the relative magnitudes of the principal moments of inertia.

Coordinates of the Rigid Rotor

Different branches of physics and engineering use different coordinates for the description of the kinematics of a rigid rotor. In molecular physics Euler angles are used almost exclusively. In quantum mechanical applications it is advantageous to use Euler angles in a convention that is a simple extension of the physical convention of spherical polar coordinates.

The first step is the attachment of a right-handed orthonormal frame (3-dimensional system of orthogonal axes) to the rotor (a body-fixed frame). This frame can be attached arbitrarily to the body, but often one uses the principal axes frame—the normalized eigenvectors of the inertia tensor, which always can be chosen orthonormal, since the tensor is symmetric. When the rotor possesses a symmetry-axis, it usually coincides with one of the principal axes. It is convenient to choose as body-fixed z-axis the highest-order symmetry axis.

One starts by aligning the body-fixed frame with a space-fixed frame (laboratory axes), so that the body-fixed x, y, and z axes coincide with the space-fixed X, Y, and Z axis. Secondly, the body and its frame are rotated actively over a positive angle α around the z-axis (by the right-hand rule), which moves the y- to the y'-axis. Thirdly, one rotates the body and its frame over a positive angle β around the y'-axis. The z-axis of the body-fixed frame has after these two rotations the longitudinal angle α (commonly designated by φ) and the colatitude angle β (commonly designated by θ), both with respect to the space-fixed frame. If the rotor were cylindrical symmetric around its z-axis, like the linear rigid rotor, its orientation in space would be unambiguously specified at this point.

If the body lacks cylinder (axial) symmetry, a last rotation around its z-axis (which has polar coordinates β and α) is necessary to specify its orientation completely. Traditionally the last rotation angle is called γ.

The convention for Euler angles is known as the $z'' - y' - z$ convention; it can be shown that it is equivalent to the $z - y - z$ convention in which the order of rotations is reversed.

The total matrix of the three consecutive rotations is the product:

$$\mathbf{R}(\alpha,\beta,\gamma)=\begin{pmatrix}\cos\alpha & -\sin\alpha & 0\\ \sin\alpha & \cos\alpha & 0\\ 0 & 0 & 1\end{pmatrix}\begin{pmatrix}\cos\beta & 0 & \sin\beta\\ 0 & 1 & 0\\ -\sin\beta & 0 & \cos\beta\end{pmatrix}\begin{pmatrix}\cos\gamma & -\sin\gamma & 0\\ \sin\gamma & \cos\gamma & 0\\ 0 & 0 & 1\end{pmatrix}$$

Let $\mathbf{r}(0)$ be the coordinate vector of an arbitrary point $\mathcal{P}$ in the body with respect to the body-fixed frame. The elements of $\mathbf{r}(0)$ are the 'body-fixed coordinates' of $\mathcal{P}$. Initially $\mathbf{r}(0)$ is also the space-fixed coordinate vector of $\mathcal{P}$. Upon rotation of the body, the body-fixed coordinates of $\mathcal{P}$ do not change, but the space-fixed coordinate vector of $\mathcal{P}$ becomes,

$$\mathbf{r}(\alpha,\beta,\gamma)=\mathbf{R}(\alpha,\beta,\gamma)\mathbf{r}(0)$$

In particular, if $\mathcal{P}$ is initially on the space-fixed Z-axis, it has the space-fixed coordinates,

$$\mathbf{R}(\alpha,\beta,\gamma)\begin{pmatrix}0\\0\\r\end{pmatrix}=\begin{pmatrix}r\cos\alpha\sin\beta\\r\sin\alpha\sin\beta\\r\cos\beta\end{pmatrix},$$

which shows the correspondence with the spherical polar coordinates (in the physical convention).

Knowledge of the Euler angles as function of time t and the initial coordinates $\mathbf{r}(0)$ determine the kinematics of the rigid rotor.

Classical Kinetic Energy

The following text forms a generalization of the well-known special case of the rotational energy of an object that rotates around *one* axis.

It will be assumed from here on that the body-fixed frame is a principal axes frame; it diagonalizes the instantaneous inertia tensor $\mathbf{I}(t)$ (expressed with respect to the space-fixed frame), i.e.,

$$\mathbf{R}(\alpha,\beta,\gamma)^{-1}\,\mathbf{I}(t)\,\mathbf{R}(\alpha,\beta,\gamma)=\mathbf{I}(0)\quad\text{with}\quad\mathbf{I}(0)=\begin{pmatrix}I_1&0&0\\0&I_2&0\\0&0&I_3\end{pmatrix},$$

Where, the Euler angles are time-dependent and in fact determine the time dependence of $\mathbf{I}(t)$ by the inverse of this equation. This notation implies that at $t=0$ the Euler angles are zero, so that at $t=0$ the body-fixed frame coincides with the space-fixed frame.

The classical kinetic energy T of the rigid rotor can be expressed in different ways:

- As a function of angular velocity.
- In Lagrangian form.
- As a function of angular momentum.
- In Hamiltonian form.

Angular Velocity Form

As a function of angular velocity T reads,

$$T=\frac{1}{2}\left[I_1\omega_x^2+I_2\omega_y^2+I_3\omega_z^2\right]$$

With,

$$\begin{pmatrix} \omega_x \\ \omega_y \\ \omega_z \end{pmatrix} = \begin{pmatrix} -\sin\beta\cos\gamma & \sin\gamma & 0 \\ \sin\beta\sin\gamma & \cos\gamma & 0 \\ \cos\beta & 0 & 1 \end{pmatrix} \begin{pmatrix} \dot{\alpha} \\ \dot{\beta} \\ \dot{\gamma} \end{pmatrix}.$$

The vector $\omega = (\omega_x, \omega_y, \omega_z)$ contains the components of the angular velocity of the rotor expressed with respect to the body-fixed frame. It can be shown that ω is *not* the time derivative of any vector, in contrast to the usual definition of velocity. The dots over the time-dependent Euler angles indicate time derivatives. The angular velocity satisfies equations of motion known as Euler's equations (with zero applied torque, since by assumption the rotor is in field-free space).

Lagrange Form

Backsubstitution of the expression of ω into T gives the kinetic energy in Lagrange form (as a function of the time derivatives of the Euler angles). In matrix-vector notation,

$$2T = \begin{pmatrix} \dot{\alpha} & \dot{\beta} & \dot{\gamma} \end{pmatrix} \mathbf{g} \begin{pmatrix} \dot{\alpha} \\ \dot{\beta} \\ \dot{\gamma} \end{pmatrix},$$

where $\mathbf{g}$ is the metric tensor expressed in Euler angles—a non-orthogonal system of curvilinear coordinates:

$$\mathbf{g} = \begin{pmatrix} I_1 \sin^2\beta\cos^2\gamma + I_2 \sin^2\beta\sin^2\gamma + I_3\cos^2\beta & (I_2 - I_1)\sin\beta\sin\gamma\cos\gamma & I_3\cos\beta \\ (I_2 - I_1)\sin\beta\sin\gamma\cos\gamma & I_1\sin^2\gamma + I_2\cos^2\gamma & 0 \\ I_3\cos\beta & 0 & I_3 \end{pmatrix}.$$

Angular Momentum Form

Often the kinetic energy is written as a function of the angular momentum $\mathbf{L}$ of the rigid rotor. With respect to the body-fixed frame it has the components L_i, and can be shown to be related to the angular velocity,

$$\mathbf{L} = \mathbf{I}(0)\,\omega \quad \text{or} \quad L_i = \frac{\partial T}{\partial \omega_i},\ i = x, y, z.$$

This angular momentum is a conserved (time-independent) quantity if viewed from a stationary space-fixed frame. Since the body-fixed frame moves (depends on time) the components L_i are *not* time independent. If we were to represent $\mathbf{L}$ with respect to the stationary space-fixed frame, we would find time independent expressions for its components.

The kinetic energy is expressed in terms of the angular momentum by:

$$T = \frac{1}{2}\left[\frac{L_x^2}{I_1} + \frac{L_y^2}{I_2} + \frac{L_z^2}{I_3}\right].$$

Hamilton Form

The Hamilton form of the kinetic energy is written in terms of generalized momenta,

$$\begin{pmatrix} p_\alpha \\ p_\beta \\ p_\gamma \end{pmatrix} \stackrel{\text{def}}{=} \begin{pmatrix} \partial T / \partial \dot{\alpha} \\ \partial T / \partial \dot{\beta} \\ \partial T / \partial \dot{\gamma} \end{pmatrix} = \mathbf{g} \begin{pmatrix} \dot{\alpha} \\ \dot{\beta} \\ \dot{\gamma} \end{pmatrix},$$

where it is used that the **g** is symmetric. In Hamilton form the kinetic energy is,

$$2T = \begin{pmatrix} p_\alpha & p_\beta & p_\gamma \end{pmatrix} \mathbf{g}^{-1} \begin{pmatrix} p_\alpha \\ p_\beta \\ p_\gamma \end{pmatrix},$$

with the inverse metric tensor given by:

$$\begin{pmatrix} \frac{\cos^2\gamma}{I_1} + \frac{\sin^2\gamma}{I_2} & \left(\frac{1}{I_2} - \frac{1}{I_1}\right)\sin\beta\sin\gamma\cos\gamma & -\frac{\cos\beta\cos^2\gamma}{I_1} - \frac{\cos\beta\sin^2\gamma}{I_2} \\ \left(\frac{1}{I_2} - \frac{1}{I_1}\right)\sin\beta\sin\gamma\cos\gamma & \frac{\sin^2\beta\sin^2\gamma}{I_1} + \frac{\sin^2\beta\cos^2\gamma}{I_2} & \left(\frac{1}{I_1} - \frac{1}{I_2}\right)\sin\beta\cos\beta\sin\gamma\cos\gamma \\ -\frac{\cos\beta\cos^2\gamma}{I_1} - \frac{\cos\beta\sin^2\gamma}{I_2} & \left(\frac{1}{I_1} - \frac{1}{I_2}\right)\sin\beta\cos\beta\sin\gamma\cos\gamma & \frac{\cos^2\beta\cos^2\gamma}{I_1} + \frac{\cos^2\beta\sin^2\gamma}{I_2} + \frac{\sin^2\beta}{I_3} \end{pmatrix}.$$

This inverse tensor is needed to obtain the Laplace-Beltrami operator, which (multiplied by $-\hbar^2$) gives the quantum mechanical energy operator of the rigid rotor.

The classical Hamiltonian given above can be rewritten to the following expression, which is needed in the phase integral arising in the classical statistical mechanics of rigid rotors,

$$\begin{aligned} T \quad = \quad & \frac{1}{2I_1\sin^2\beta}\left((p_\alpha - p_\gamma\cos\beta)\cos\gamma - p_\beta\sin\beta\sin\gamma\right)^2 \\ & + \frac{1}{2I_2\sin^2\beta}\left((p_\alpha - p_\gamma\cos\beta)\sin\gamma + p_\beta\sin\beta\cos\gamma\right)^2 + \frac{p_\gamma^2}{2I_3}. \end{aligned}$$

Quantum Mechanical Rigid Rotor

As usual quantization is performed by the replacement of the generalized momenta by operators that give first derivatives with respect to its canonically conjugate variables (positions). Thus,

$$p_\alpha \rightarrow -i\hbar\frac{\partial}{\partial\alpha}$$

and similarly for p_β and p_γ. It is remarkable that this rule replaces the fairly complicated function p_α of all three Euler angles, time derivatives of Euler angles, and inertia moments (characterizing the rigid rotor) by a simple differential operator that does not depend on time or inertia moments and differentiates to one Euler angle only.

The quantization rule is sufficient to obtain the operators that correspond with the classical angular momenta. There are two kinds: space-fixed and body-fixed angular momentum operators. Both are vector operators, i.e., both have three components that transform as vector components among themselves upon rotation of the space-fixed and the body-fixed frame, respectively. The explicit form of the rigid rotor angular momentum operators is given here (but beware, they must be multiplied with $\hbar$). The body-fixed angular momentum operators are written as $\widehat{\mathcal{P}}_i$. They satisfy anomalous commutation relations.

The quantization rule is *not* sufficient to obtain the kinetic energy operator from the classical Hamiltonian. Since classically p_β commutes with $\cos\beta$ and $\sin\beta$ and the inverses of these functions, the position of these trigonometric functions in the classical Hamiltonian is arbitrary. After quantization the commutation does no longer hold and the order of operators and functions in the Hamiltonian (energy operator) becomes a point of concern. Podolsky proposed in 1928 that the Laplace-Beltrami operator (times $-\frac{1}{2}\hbar^2$) has the appropriate form for the quantum mechanical kinetic energy operator. This operator has the general form (summation convention: sum over repeated indices—in this case over the three Euler angles $q^1, q^2, q^3 \equiv \alpha, \beta, \gamma$):

$$\hat{H} = -\tfrac{\hbar^2}{2} |g|^{-1/2} \frac{\partial}{\partial q^i} |g|^{1/2} g^{ij} \frac{\partial}{\partial q^j},$$

where $|g|$ is the determinant of the g-tensor:

$$|g| = I_1 I_2 I_3 \sin^2\beta \quad \text{and} \quad g^{ij} = (\mathbf{g}^{-1})_{ij}.$$

Given the inverse of the metric tensor above, the explicit form of the kinetic energy operator in terms of Euler angles follows by simple substitution. (The corresponding eigenvalue equation gives the Schrödinger equation for the rigid rotor in the form that it was solved for the first time by Kronig and Rabi (for the special case of the symmetric rotor). This is one of the few cases where the Schrödinger equation can be solved analytically. All these cases were solved within a year of the formulation of the Schrödinger equation).

Nowadays it is common to proceed as follows. It can be shown that $\hat{H}$ can be expressed in body-fixed angular momentum operators (in this proof one must carefully commute differential operators with trigonometric functions). The result has the same appearance as the classical formula expressed in body-fixed coordinates,

$$\hat{H} = \tfrac{1}{2}\left[\frac{\mathcal{P}_x^2}{I_1} + \frac{\mathcal{P}_y^2}{I_2} + \frac{\mathcal{P}_z^2}{I_3}\right].$$

The action of the $\widehat{\mathcal{P}}_i$ on the Wigner D-matrix is simple. In particular

$$\mathcal{P}^2 D^j_{m'm}(\alpha,\beta,\gamma)^* = \hbar^2 j(j+1) D^j_{m'm}(\alpha,\beta,\gamma)^* \quad \text{with} \quad \mathcal{P}^2 = \mathcal{P}_x^2 + \mathcal{P}_y^2 + \mathcal{P}_z^2,$$

so that the Schrödinger equation for the spherical rotor ($I = I_1 = I_2 = I_3$) is solved with the $(2j+1)^2$ degenerate energy equal to $\frac{\hbar^2 j(j+1)}{2I}$.

The symmetric top (= symmetric rotor) is characterized by $I_1 = I_2$. It is a *prolate* (cigar shaped) top if $I_3 < I_1 = I_2$. In the latter case we write the Hamiltonian as:

$$\hat{H} = \frac{1}{2}\left[\frac{\mathcal{P}^2}{I_1} + \mathcal{P}_z^2\left(\frac{1}{I_3} - \frac{1}{I_1}\right)\right],$$

and use that:

$$\mathcal{P}_z^2\, D^j_{mk}(\alpha,\beta,\gamma)^* = \hbar^2 k^2\, D^j_{mk}(\alpha,\beta,\gamma)^*.$$

Hence,

$$\hat{H}\, D^j_{mk}(\alpha,\beta,\gamma)^* = E_{jk} D^j_{mk}(\alpha,\beta,\gamma)^* \quad \text{with} \quad E_{jk}/\hbar^2 = \frac{j(j+1)}{2I_1} + k^2\left(\frac{1}{2I_3} - \frac{1}{2I_1}\right).$$

The eigenvalue E_{j0} is $2j+1$-fold degenerate, for all eigenfunctions with $m = -j, -j+1, \ldots, j$ have the same eigenvalue. The energies with |k| > 0 are $2(2j+1)$-fold degenerate. This exact solution of the Schrödinger equation of the symmetric top was first found in 1927.

The asymmetric top problem ($I_1 \neq I_2 \neq I_3$) is not exactly soluble.

Direct Experimental Observation of Molecular Rotations

For a long time, molecular rotations could not be directly observed experimentally. Only measurement techniques with atomic resolution made it possible to detect the rotation of a single molecule. At low temperatures, the rotations of molecules (or part thereof) can be frozen. This could be directly visualized by Scanning tunneling microscopy i.e., the stabilization could be explained at higher temperatures by the rotational entropy. The direct observation of rotational excitation at single molecule level was achieved recently using inelastic electron tunneling spectroscopy with the scanning tunneling microscope. The rotational excitation of molecular hydrogen and its isotopes were detected.

Euler's Equations

The fundamental equation of motion of a rotating body,

$$\mathbf{T} = \frac{d\mathbf{L}}{dt},$$

is only valid in an *inertial* frame. However, we have seen that **L** is most simply expressed in a frame of reference whose axes are aligned along the principal axes of rotation of the body. Such a frame of reference *rotates* with the body, and is, therefore, *non-inertial*. Thus, it is helpful to define *two* Cartesian coordinate systems, with the same origins. The first, with coordinates x, y, z, is a fixed inertial frame-let us denote this the *fixed frame*. The second, with coordinates x', y', z', co-rotates with the body in such a manner that the x'-, y'-, and z'-axes are always pointing along its principal axes of rotation--we shall refer to this as the *body frame*. Since the body frame co-rotates with the body, its instantaneous angular velocity is the same as that of the body. Hence, it follows that,

$$\frac{d\mathbf{L}}{dt} = \frac{d\mathbf{L}}{dt'} + \omega \times \mathbf{L}$$

Here, d/dt is the time derivative in the fixed frame, and d/dt' the time derivative in the body frame. Combining Equations $\mathbf{T} = \frac{d\mathbf{L}}{dt}$ and $\frac{d\mathbf{L}}{dt} = \frac{d\mathbf{L}}{dt'} + \omega \times \mathbf{L}$, we obtain,

$$\mathbf{T} = \frac{d\mathbf{L}}{dt'} + \omega \times \mathbf{L}.$$

Now, in the body frame let $T = (T_{x'}, T_{y'}, T_{z'})$ and $\omega = (\omega_{x'}, \omega_{y'}, \omega_{z'})$. It follows that $L = (I_{x'x'}\,\omega_{x'}, L_{y'y'}\,\omega_{y'}, I_{z'z'}\,\omega_{z'})$, where $I_{x'x'}$, $I_{y'y'}$ and $I_{z'z'}$ are the principal moments of inertia. Hence, in the body frame, the components of equation $\mathbf{T} = \frac{d\mathbf{L}}{dt'} + \omega \times \mathbf{L}$ yield,

$$T_{x'} = I_{x'x'}\,\dot{\omega}_{x'} - (I_{y'y'} - I_{z'z'})\,\omega_{y'}\,\omega_{z'},$$
$$T_{y'} = I_{y'y'}\,\dot{\omega}_{y'} - (I_{z'z'} - I_{x'x'})\,\omega_{z'}\,\omega_{x'},$$
$$T_{z'} = I_{z'z'}\,\dot{\omega}_{z'} - (I_{x'x'} - I_{y'y'})\,\omega_{x'}\,\omega_{y'},$$

where $= d/dt'$. Here, we have made use of the fact that the moments of inertia of a rigid body are *constant* in time in the co-rotating body frame. The above equations are known as *Euler's equations.*

Consider a rigid body which is constrained to rotate about a fixed axis with *constant* angular velocity. It follows that $\dot{\omega}_{x'} = \dot{\omega}_{y'} = \dot{\omega}_{z'} = 0$. Hence, Euler's equations, $T_{x'} = -(I_{y'y'} - I_{z'z'})\,\omega_{y'}\,\omega_{z'}$, $T_{y'} = -(I_{z'z'} - I_{x'x'})\,\omega_{z'}\,\omega_{x'}$, $T_{z'} = -(I_{x'x'} - I_{y'y'})\,\omega_{x'}\,\omega_{y'}$, reduce to,

$$T_{x'} = -(I_{y'y'} - I_{z'z'})\,\omega_{y'}\,\omega_{z'},$$
$$T_{y'} = -(I_{z'z'} - I_{x'x'})\,\omega_{z'}\,\omega_{x'},$$
$$T_{z'} = -(I_{x'x'} - I_{y'y'})\,\omega_{x'}\,\omega_{y'},$$

These equations specify the components of the steady (in the body frame) torque exerted on the body by the constraining supports. The steady (in the body frame) angular momentum is written:

$$L = (I_{x'x'}\,\omega_{x'}, I_{y'y'}\,\omega_{y'}, I_{z'z'}\,\omega_{z'}).$$

It is easily demonstrated that $T = \omega \times L$. Hence, the torque is perpendicular to both the angular velocity and the angular momentum vectors. Note that if the axis of rotation is a principal axis then two of the three components of ω are zero (in the body frame). It follows from Equations $T_{x'} = -(I_{y'y'} - I_{z'z'})\,\omega_{y'}\,\omega_{z'}$, $T_{y'} = -(I_{z'z'} - I_{x'x'})\,\omega_{z'}\,\omega_{x'}$, $T_{z'} = -(I_{x'x'} - I_{y'y'})\,\omega_{x'}\,\omega_{y'}$, that all three components of the torque are zero. In other words, *zero* external torque is required to make the body rotate steadily about a *principal axis*.

Suppose that the body is freely rotating: i.e., there are no external torques. Furthermore, let the body be rotationally symmetric about the z'-axis. It follows that $I_{x'x'} = I_{y'y'} = I_{\perp}$. Likewise, we can write $I_{z'z'} = I_{\parallel}$. In general, however, $I_{\perp} \neq I_{\parallel}$. Thus, Euler's equations yield,

$$I_{\perp} = \frac{d\omega_{x'}}{dt'} + \left(I_{\parallel} - I_{\perp}\right)\omega_{z'}\,\omega_{y'} = 0$$

$$I_{\perp} = \frac{d\omega_{x'}}{dt'} + \left(I_{\parallel} - I_{\perp}\right)\omega_{z'}\,\omega_{x'} = 0$$

$$\frac{d\omega_{z'}}{dt'} = 0$$

Clearly, $\omega_{z'}$ is a constant of the motion. Equation $I_{\perp} = \frac{d\omega_{x'}}{dt'} + \left(I_{\parallel} - I_{\perp}\right)\omega_{z'}\,\omega_{y'} = 0$ and $I_{\perp} = \frac{d\omega_{x'}}{dt'} + \left(I_{\parallel} - I_{\perp}\right)\omega_{z'}\,\omega_{x'} = 0$ can be written:

$$\frac{d\omega_{x'}}{dt'} + \Omega\omega_{y'} = 0$$

$$\frac{d\omega_{x'}}{dt'} + \Omega\omega_{x'} = 0$$

where $\Omega = \left(I_{\parallel} / I_{\perp} - 1\right)\omega_{z'}$. As is easily demonstrated, the solution to the above equations is:

$$\omega_{x'} = \omega_{\perp}\cos\left(\Omega t'\right),$$

$$\omega_{y'} = \omega_{\perp}\cos\left(\Omega t'\right),$$

where $\omega_{\perp}$ is a constant. Thus, the projection of the angular velocity vector onto the x'-y' plane has the fixed length $\omega_{\perp}$, and rotates steadily about the z'-axis with angular velocity Ω. It follows that the length of the angular velocity vector, $\omega = \left(\omega_{x'}^2 + \omega_{y'}^2 + \omega_{z'}^2\right)^{1/2}$, is a constant of the motion. Clearly, the angular velocity vector makes some constant angle, α, with the -axis, which implies that $\omega_{z'} = \omega\cos\alpha$ and $\omega_{\perp'} = \omega\sin\alpha$. Hence, the components of the angular velocity vector are:

$$\omega_{x'} = \omega\sin\alpha\cos\left(\Omega t'\right),$$

$$\omega_{y'} = \omega\sin\alpha\cos\left(\Omega t'\right),$$

$$\omega_{z'} = \omega\cos\alpha,$$

Where,

$$\Omega = \omega\cos\alpha\left(\frac{I_{\parallel}}{I_{\perp}} - 1\right).$$

We conclude that, in the body frame, the angular velocity vector *precesses* about the symmetry axis (*i.e.*, the z'-axis) with the angular frequency Ω. Now, the components of the angular momentum vector are:

$$L_{x'} = I_{\perp}\omega\sin\alpha\cos(\Omega t'),$$
$$L_{y'} = I_{\perp}\omega\sin\alpha\cos(\Omega t'),$$
$$L_{z'} = I_{\parallel}\omega\cos\alpha.$$

Thus, in the body frame, the angular momentum vector is also of constant length, and precesses about the symmetry axis with the angular frequency Ω. Furthermore, the angular momentum vector makes a constant angle θ with the symmetry axis, where:

$$\tan\theta = \frac{I_{\perp}}{I_{\parallel}}\tan\alpha.$$

Note that the angular momentum vector, the angular velocity vector, and the symmetry axis all lie in the *same plane*: *i.e.*, $e_{z'} \cdot L \times \omega = 0$, as can easily be verified. Moreover, the angular momentum vector lies between the angular velocity vector and the symmetry axis (*i.e.*, $\theta < \alpha$) for a flattened (or oblate) body (*i.e.*, $I_{\perp} < I_{\parallel}$), whereas the angular velocity vector lies between the angular momentum vector and the symmetry axis (*i.e.*, $\theta > \alpha$) for an elongated (or prolate) body (*i.e.*, $I_{\perp} > I_{\parallel}$).

Let us now consider the most general motion of a freely rotating *asymmetric* rigid body, as seen in the body frame. Since a freely rotating body experiences no external torques, its angular momentum vector L is a constant of the motion in the inertial fixed frame. In general, the direction of this vector varies with time in the non-inertial body frame, but its length remains fixed. This can be seen from Equation $\frac{d\mathbf{L}}{dt} = \frac{d\mathbf{L}}{dt'} + \omega \times \mathbf{L}$: if $dL/dt = 0$ then the scalar product of this equation with L implies that $dL^2/dt' = 0$. It follows from Equation $L = (I_{x'x'}\,\omega_{x'}, I_{y'y'}\,\omega_{y'}, I_{z'z'}\,\omega_{z'})$ that:

$$L^2 = I_{x'x'}^2\,\omega_{x'}^2 + I_{y'y'}^2\,\omega_{y'}^2 + I_{z'z'}^2\,\omega_{z'}^2 = \text{constant}.$$

The above constraint can also be derived directly from Euler's equations,

$$T_{y'} = I_{y'y'}\,\dot{\omega}_{y'} - (I_{z'z'} - I_{x'x'})\,\omega_{z'}\,\omega_{x'},$$
$$T_{z'} = I_{z'z'}\,\dot{\omega}_{z'} - (I_{x'x'} - I_{y'y'})\,\omega_{x'}\,\omega_{y'},$$
$$T_{x'} = I_{x'x'}\,\dot{\omega}_{x'} - (I_{y'y'} - I_{z'z'})\,\omega_{y'}\,\omega_{z'},$$

by setting $T_{x'} = T_{y'} = T_{z'} = 0$. A freely rotating body subject to no external torques clearly has a constant rotational kinetic energy. Hence, from equation,

$$\omega \cdot \mathbf{L} = I_{x'x'}\,\omega_{y'}^2 + I_{y'y'}\,\omega_{y'}^2 + I_{z'z'}\,\omega_{z'}^2 = \text{constant}.$$

This constraint can also be derived directly from Euler's equations. We conclude that, in the body frame, the components of ω must simultaneously satisfy the two constraints $L^2 = I_{x'x'}^2\,\omega_{x'}^2 + I_{y'y'}^2\,\omega_{y'}^2 + I_{z'z'}^2\,\omega_{z'}^2 = \text{constant}$ and $\omega \cdot \mathbf{L} = I_{x'x'}\,\omega_{y'}^2 + I_{y'y'}\,\omega_{y'}^2 + I_{z'z'}\,\omega_{z'}^2 = \text{constant}$. These constraints are the equations of two ellipsoids whose principal axes coincide with the principal axes of the body, and whose principal radii are in the ratio $I_{x'x'}^{-1} : I_{y'y'}^{-1} : I_{z'z'}^{-1}$ and $I_{x'x'}^{-1/2} : I_{y'y'}^{-1/2} : I_{z'z'}^{-1/2}$, respectively. In general, the intersection of these two ellipsoids is a *closed curve*. Hence, we conclude that the most general motion of a freely rotating asymmetric body, as seen in the body frame, is a form of irregular *precession* in which the tip of the angular velocity vector ω periodically traces out the aforementioned closed curve. It is easily demonstrated that the tip of the angular momentum vector L periodically traces out a different closed curve.

References

- Rotational-motion-of-a-rigid-body, physics-classical-mechanics: cliffsnotes.com, Retrieved 02 June, 2019
- Tenenbaum, RA (2004). Fundamentals of Applied Dynamics. Springer. ISBN 0-387-00887-X
- Bigoni, D. Nonlinear Solid Mechanics: Bifurcation Theory and Material Instability. Cambridge University Press, 2012 . ISBN 9781107025417
- Goldstein, H.; Poole, C. P.; Safko, J. L. (2001). Classical Mechanics (Third ed.). San Francisco: Addison Wesley Publishing Company. ISBN 0-201-65702-3
- Thomas Waldmann; Jens Klein; Harry E. Hoster; R. Jürgen Behm (2012), “Stabilization of Large Adsorbates by Rotational Entropy: A Time-Resolved Variable-Temperature STM Study”, ChemPhysChem (in German), 14, pp. 162–169, doi:10.1002/cphc.201200531, PMID 23047526

PERMISSIONS

All chapters in this book are published with permission under the Creative Commons Attribution Share Alike License or equivalent. Every chapter published in this book has been scrutinized by our experts. Their significance has been extensively debated. The topics covered herein carry significant information for a comprehensive understanding. They may even be implemented as practical applications or may be referred to as a beginning point for further studies.

We would like to thank the editorial team for lending their expertise to make the book truly unique. They have played a crucial role in the development of this book. Without their invaluable contributions this book wouldn't have been possible. They have made vital efforts to compile up to date information on the varied aspects of this subject to make this book a valuable addition to the collection of many professionals and students.

This book was conceptualized with the vision of imparting up-to-date and integrated information in this field. To ensure the same, a matchless editorial board was set up. Every individual on the board went through rigorous rounds of assessment to prove their worth. After which they invested a large part of their time researching and compiling the most relevant data for our readers.

The editorial board has been involved in producing this book since its inception. They have spent rigorous hours researching and exploring the diverse topics which have resulted in the successful publishing of this book. They have passed on their knowledge of decades through this book. To expedite this challenging task, the publisher supported the team at every step. A small team of assistant editors was also appointed to further simplify the editing procedure and attain best results for the readers.

Apart from the editorial board, the designing team has also invested a significant amount of their time in understanding the subject and creating the most relevant covers. They scrutinized every image to scout for the most suitable representation of the subject and create an appropriate cover for the book.

The publishing team has been an ardent support to the editorial, designing and production team. Their endless efforts to recruit the best for this project, has resulted in the accomplishment of this book. They are a veteran in the field of academics and their pool of knowledge is as vast as their experience in printing. Their expertise and guidance has proved useful at every step. Their uncompromising quality standards have made this book an exceptional effort. Their encouragement from time to time has been an inspiration for everyone.

The publisher and the editorial board hope that this book will prove to be a valuable piece of knowledge for students, practitioners and scholars across the globe.

INDEX